Pressure Pipeline Design for Water and Wastewater

Prepared by the Committee on Pipeline Planning
of the Pipeline Division
of the American Society of Civil Engineers

ASCE 1852

Published by the
American Society of Civil Engineers
345 East 47th Street
New York, New York 10017-2398

ABSTRACT

This committee report, Pressure Pipeline Design for Water and Wastewater, is a revision of the 1975 ASCE publication Pipeline Design for Water and Wastewater. This committee report presents updated information on all aspects of transporting fluids through pipelines such as: 1) Accepted flow formulas, 2) surge control in pumped pipelines, 3) main pipeline design, 4) pipe materials, and 5) pipeline economics. In addition, a new section on seismic design prepared by the ASCE Technical Council on Lifeline Earthquake Engineering, Water and Sewage Committee, was added. This report will be of assistance to engineers who are occasionally concerned with pipeline design as well as the specialists in pipeline design.

Library of Congress Cataloging-in-Publication Data

Pressure pipeline design for water and wastewater / prepared by the Committee on Pipeline Planning of the Pipeline Division of the American Society of Civil Engineers. — 2nd ed.

p.cm.
Includes index.
ISBN 0-87262-931-7

1. Water-pipes—Design and construction. 2. Sewer-pipe—Design and construction. 3. Earthquake resistant design. I. American Society of Civil Engineers. Committee on Pipeline Planning.
TD491.P743 1992
628.1'5—dc20 92-41622
CIP

The material presented in this publication has been prepared in accordance with generally recognized engineering principals and practices, and is for general information only. This information should not be used without first securing competent advice with respect to its suitability for any general or specific application.

The contents of this publication are not intended to be and should not be construed to be a standard of the American Society of Civil Engineers (ASCE) and are not intended for use as a reference in purchase specifications, contracts, regulations, statutes, or any other legal document.

No reference made in publication to any specific method, product, process, or service constitutes or implies an endorsement, recommendation, or warranty thereof by ASCE.

ASCE makes no representation or warranty of any kind, whether express or implied, concerning the accuracy, completeness, suitability or utility of any information, apparatus, product or process discussed in this publication, and assumes no liability therefor.

Anyone utilizing this information assumes all liability arising from such use, including but not limited to infringement of any patent or patents.

Library of Congress Catalog Card No: 92-41622
ISBN 0-87262-931-7
Manufactured in the United States of America.

TABLE OF CONTENTS

TABLE OF CONTENTS (Continued)

TABLE OF CONTENTS (Continued)

TABLE OF CONTENTS (Continued)

APPENDICES

LIST OF TABLES

LIST OF FIGURES

Page

LIST OF FIGURES (Continued)

INTRODUCTION

In 1964, the Committee on Pipeline Planning formed a task Committee on Engineering Practice in the Design of Pipelines (EPDP), which undertook to produce a report of the most widely accepted flow formulas and appurtenant design data to the end that not only the pipeline engineer, but also the engineer not ordinarily concerned with pipeline design on a day-to-day basis might turn to for information.

The result was a publication called, "Pipeline Design For Water And Wastewater", published by American Society of Civil Engineers in 1975.

In 1989, a subcommittee of the Pipeline Planning Committee was formed to review and update the 1975 manual. Each section was updated. A new section, Seismic Resistant Pipeline Design prepared by the ASCE Technical Council on Lifeline Earthquake Engineering, Water and Sewage Committee, was added. This manual is the results of that effort.

As in the 1975 manual, discussions of structural loading on buried pipes have not been included because this broad subject is not unique to pressure pipelines. An ASCE publication, "Gravity Sanitary Sewer Design and Construction", Manual No. 60, is recommended as a basic reference. This manual covers rigid and flexible pipe materials while proceedings papers from the annual Pipeline Division Specialty Conferences will include the latest information on loading research.

Many people have contributed to this work by adding materials, rewriting sections, reviewing material and editing manuscripts. They are the following: Deon T. Fowles, Malcolm Stephens, Gary Z. Watters, Robert T. Bates, Richard Fuerst, Dr. Leon Ru-Liang Wang, Don Ballantyne, Kenneth D. Kerri, Joe Richard, Stan Spaulding, Dennis E. Bauer, Jack Jolley, Robert J. Bailey, John A. Anderson, Richard Bonds, Joe Barber, Ben C. Yen, Marshall Marriott, Rick Turkopp, Robert P. Walker, Stanley A. Mruk, Timothy B. Kennedy, A. J. Hoffman, and Duane B. Ford.

Thanks are extended to the many secretaries who have typed these sections. Especially to Sherri Hodges who did the final typing and preparation of these pages for publication.

Appreciation for the efforts of everyone involved is hereby extended.

SECTION 1 - WATER AND WASTEWATER CHARACTERISTICS

1.1 Water

Water in the liquid state can be considered for all practical purposes to have a specific gravity of unity and to be incompressible. Raw waters vary widely from highly corrosive to non-corrosive, and some contain dissolved minerals which deposit on the inside of pipelines in scale formations. Water from surface sources may contain small quantities of suspended matter, organic substances, or organisms as well as dissolved solids. The temperature of surface waters fluctuates with climatic conditions. Groundwaters frequently contain dissolved minerals and sometimes pesticides, gases, tastes, and odors. Groundwater temperatures normally approach the average annual temperature in the region.

When raw waters contain excessive quantities of sediments or other impurities, treatment is required. Usually excessive corrosive waters are treated to reduce this quality in conjunction with any other types of treatment required. The temperature of treated water is generally the same or slightly higher than raw water due to the period of detention in the treatment plant subject to seasonal higher temperatures or winter heating. Buried pipelines, however, tend to bring water temperatures to near the average annual or seasonal temperature of the region, depending on depth of burial.

High temperatures accelerate corrosive actions but decrease the viscosity of the water. Some properties of water are listed in Table 1.

TABLE 1 - PROPERTIES OF WATER

Temperature °F	Specific Gravity	Specific Weight lb/ft^3	Absolute Viscosity (μ) lb sec/ft^2	Kinematic Viscosity (ν) ft^2/sec
32	0.9999	62.42	3.746×10^{-5}	1.931×10^{-5}
39.2	1.0000	62.427	3.274 x "	1.687 "
50	0.9997	62.41	2.735 x "	1.410 "
60	0.9990	62.37	2.359 x "	1.217 "
70	0.9980	62.30	2.050 x "	1.059 "
80	0.9966	62.22	1.799 x "	.0930 "
100	0.9931	62.00	1.424 x "	0.739 "
Sea Water		64		

1.2 Wastewater

Wastewater is the used and spent waters of a community or industry. The former has generally been called sanitary sewage, the latter industrial wastes. The physical characteristics of wastewater, while reflecting the source water, are primarily those which result from the changes it has undergone during use. Thus, the temperature of sanitary sewage tends to be more uniform the year around, usually varying between 50°F (10°C) and 70°F (21°C). From 300 mg/l to 500 mg/l of dissolved solids and 200 mg/l to 400 mg/l suspended solids including grease will have been added during use. For all practical purposes however, sanitary sewage has the same physical characteristics which affect hydraulic design as water of the same temperature.

Unlike sanitary sewage, industrial wastewaters can vary considerably in temperature and other physical characteristics. Dissolved and suspended solids could conceivably affect density and viscosity of the fluid. Laboratory testing is required to determine if these physical characteristics vary sufficiently from water of the same temperature to affect hydraulic design.

While the physical qualities of wastewaters affecting hydraulic design are often similar to those of water, those qualities affecting the corrosivity of the fluid can vary markedly from water. Sanitary sewage that has become septic can, under some circumstances, produce hydrogen sulfide which can be corrosive to cement and metal products. Industrial wastewaters can vary from strongly acidic to strongly alkaline, requiring special materials or linings for pipelines, pumps and other appurtenances.

This manual is concerned solely with the design of pressure pipelines, commonly known as force mains, receiving discharge from wastewater pumping stations. For the hydraulic design of wastewaters carrying suspended solids in such concentration as to affect its hydraulic characteristics, other reference material is needed.

The design of pipes carrying wastewaters by gravity flow, generally known as sewers, is fully detailed in the publication, "Design and Construction of Sanitary and Storm Sewers" (ASCE Manual on Engineering Practice No. 37, Water Pollution Control Federation Manual of Practice No. 9).

1.3 Nonpotable Water

There is a growing trend toward the use of partially treated wastewater for irrigation, and industrial uses. This is done as a means of water conservation. The pipeline systems are similar to ordinary potable water systems except that a body of standards is coming into development, intending to avoid health problems due to misuse.

Some of the pipeline system standards currently proposed include consideration of line pressure, and generally, the nonpotable system should operate at about 10 psi lower pressure than potable water systems to avoid possible contamination. It is also generally advisable to maintain a separation of 10 feet horizontal and one foot below drinking water lines for the nonpotable pipeline. At crossings of the two lines, the nonpotable should be lower by at least a foot.

Identification of nonpotable lines is of paramount concern. The pipes and service lines should be a distinctive color (purple has been suggested) and should be embossed or integrally stamped/marked CAUTION: NONPOTABLE WATER - DO NOT DRINK or installed with a purple identification tape, or a purple polyethylene vinyl wrap. Color coding should be consistent throughout the service area.

Valve boxes should be the standard concrete or fiberglass with a special triangular heavy duty cover. All valve covers should be non-interchangeable with potable water covers and with a recognizable inscription cast on the top surface.

Above ground, all exposed facilities shall be consistently color coded and marked to differential nonpotable water facilities from potable or wastewater systems.

SECTION 2 - TYPES OF FLOW

2.1 The Reynolds Number

The Reynolds Number (**R**) is a dimensionless numerical quantity used as an index to characterize the type of flow. It is equal to the product of a characteristic velocity of the system (it may be the mean, surface, or maximum velocity) and a characteristic linear dimension, such as diameter or depth, divided by the kinematic viscosity of the liquid; all expressed in consistent units. In a pipe flowing full,

$$\mathbf{R} = \frac{d\,V}{\nu} \quad \text{.....................(1)}$$

in which **R** is the dimensionless Reynolds Number, d is the inside diameter of the pipe (ft or m), V is the mean velocity of flow (fps or mps), and ν is the kinematic viscosity of the fluid (ft^2/sec or m^2/sec). The number is chiefly applicable to closed systems of flow such as pipes or conduits where there is no free water surface, or to bodies fully immersed in the fluid so the free surface need not be considered.

2.2 Laminar and Turbulent Flow

The flow of liquid in a pipe is either laminar, transitional or turbulent ("rough") and can be classified as to the numerical value of the Reynolds Number. Pipe friction coefficients f for laminar flow are independent of pipe roughness and vary only with the relative strength of the viscous and inertial forces or, the Reynolds Number. Laminar flow occurs when the Reynolds Number is less than 2,000.

Liquid flow in the range of Reynolds Numbers between 2,000 and 4,000 changes from laminar toward turbulent. Values of f are uncertain in this range. Conservative design for **R** between 2,000 and 4,000 would assume turbulent flow for selection of f.

The characteristics of turbulent flow, **R** greater than 4,000, are important since almost all flows in a water system are in this category. Piping systems are too large for the particular installation if the Reynolds Number is less than 4,000.

Flow coefficients applicable to turbulent conditions, when using hydraulic formulas with the Reynolds Number, vary with the roughness of the interior of the pipe was well as viscosity and density of the fluid. Turbulent flow is further classified as: (1) flow in smooth pipes, (2) flow in relatively rough pipe at high velocities, and (3) flow in the transition zone between these two types. Most of the normal hydraulic problems are related to flow in the last category.

In summary, the Reynolds Number provides a useful index to identify laminar or turbulent flows. The following Table 2 illustrates the relationship between the Reynolds Numbers and types of flows. The Reynolds Number for complete turbulence varies with the relative roughness of the pipe.

TYPES OF FLOW

TABLE 2 RELATIONSHIP BETWEEN REYNOLDS NUMBER AND TYPES OF FLOW

Reynolds Numbers (R)	Type of Flow
Less than 2,000	Laminar
Between 2,000 and 4,000	Unstable
Over 4,000	Transition to Turbulent

SECTION 3 - FORMULAS

3.1 Empirical and Rational Formulas

Laws applicable to laminar flow of liquids in pipes can be determined rationally. On the other hand, laws governing turbulent flow must be approximated since the phenomena of turbulence is not presently well understood. Chezy (1775) stated that the loss of head in the flow of water in conduits varied approximately as the square of velocity. Nearly a century later Darcy, Weisbach, and others accepted Chezy's hypothesis and proposed the now well-known Darcy-Weisbach pipe formula:

$$h_f = f \cdot \frac{l}{d} \cdot \frac{V^2}{2g} \qquad (2)$$

in which h_f is the energy loss (ft or m) due to friction in a length of pipe l (ft or m) and inside diameter d (ft or m), the dimensionless f is the Darcy-Weisbach friction coefficient, g is the gravitational (free fall acceleration) constant (ft/sec^2 or m/sec^2), and V is the velocity of the fluid (ft/sec or m/sec). The coefficient f depends on the value of the Reynolds number (**R**) and the value of the dimensionless number ϵ/d, representing the relative roughness of the pipe wall, in which ϵ is the equivalent roughness of the pipe wall (ft or m).

The flow of water in pipelines has been determined empirically with sufficient accuracy to meet ordinary practical requirements. The above mentioned Darcy-Weisbach formula (Equation 2) meets these practical requirements; similarly regarded are the Manning, Hazen-Williams, and Scobey formulas (see Table 3). Other experiments have shown that the effect of interior roughness is related to the pipe diameter. Rational and semi-empirical formulas in this category are identified as the Hagen-Poiseuille, Prandtl-Von Karman, Colebrook-White, and Morris. All of the formulas have the same uncertainty in the evaluation of the actual interior surface roughness of the pipe and at what rate the roughness will increase with age. The determination of these two conditions is more important than the formula used in the calculation for friction loss. For more extensive discussions on the subject of empirical formulas for flow in pipelines the reader's attention is invited to texts by H. W. King et al (Ref. 1 and 6) and S. Whitaker (Ref. 3).

TABLE 3
EMPIRICAL FLOW FORMULAS

Originator	For velocity, V	For head loss, $h_f = Sl$	For slope of energy gradient, S
Darcy, Weisbach (Fanning) English	$8.022\sqrt{\frac{dS}{f}}$	$\frac{f}{64.348}\ \frac{l}{d}\ V^2$	$\frac{0.025Q^2}{d^5} \cdot f$
Metric	$4.429\sqrt{\frac{dS}{f}}$	$\frac{f}{19.613}\ \frac{l}{d}\ V^2$	$\frac{0.083Q^2}{d^5} \cdot f$
Hazen Williams English	$1.318C_hR^{0.63}S^{0.54}$	$\left(\frac{1.816}{C_h}\right)^{1.852} \frac{l}{d^{1.167}} V^{1.852}$	$\left(\frac{2.31}{C_h}\right)^{1.852} \frac{Q^{1.852}}{d^{4.87}}$
Metric	$0.85C_hR^{0.63}S^{0.54}$	$\left(\frac{2.816}{C_h}\right)^{1.852} \frac{l}{d^{1.167}} V^{1.852}$	$\left(\frac{3.59}{C_h}\right)^{1.852} \frac{Q^{1.852}}{d^{4.87}}$
Scobey for concrete pipe English	$355C_SR^{0.625}S^{0.5}$	$\left(\frac{0.00669}{C_S}\right)^2 \frac{l}{d^{1.25}} V^2$	$\left(\frac{0.00853}{C_S}\right)^2 \frac{Q^2}{d^{5.25}}$
Metric	$227C_SR^{0.625}S^{0.5}$	$\left(\frac{0.0105}{C_S}\right)^2 \frac{l}{d^{1.25}} V^2$	$\left(\frac{0.0133}{C_S}\right)^2 \frac{Q^2}{d^{5.25}}$

In Table 3, the dimensionless S is the energy loss per unit length of pipe (h_f/l), Q is the volume rate of flow (cfs or m^3/sec), R is the hydraulic radius (ft or m), and C_h and C_s are the Hazen-Williams and Scobey friction coefficients respectively.

The Darcy-Weisbach formula, with the additional work by others with regard to the evaluation of the friction factor (f) in that formula, does permit the proper determination of all factors which affect the friction loss in a pipe, and is widely used in academic circles. On the other hand, the Hazen-Williams formulas have found almost exclusive acceptance by waterworks engineers in this country. This is due to the simplification of problem solution by the use of electronic calculations, computers, tables, nomographs and charts. In addition, the examination and flow testing of existing lines and systems using the formulas has produced a background of data of the actual value of the roughness coefficient (C_h) which allows many comparisons of test data with calculated solutions. The formula is readily adaptable to a computer program of wide scope and is used almost exclusively by computer time sharing centers as well as by most consulting engineers.

Pertinent data for application in the Hazen-Williams formula are furnished in Table 4.

The roughness coefficients (C_h) listed are applicable to the character of the pipe interiors described.

Extensive tables of friction losses using the Hazen-Williams formula may be found in Hydraulic Tables by Williams and Hazen (Ref. 4).

TABLE 4. DATA FOR THE HAZEN-WILLIAMS FORMULA

	Hazen-Williams Formula	For velocity, V feet/sec or meters/sec	For slope of hydraulic gradient, S (energy loss per unit length of pipe)
English	$V=1.318C_hR^{0.63}S^{0.54}$	$0.550C_hd^{0.63}S^{0.54}$	$\left(\frac{2.31Q}{C_hd^{2.63}}\right)^{1.852}$
Metric	$V=0.85C_hR^{0.63}S^{0.54}$	$0.355C_hd^{0.63}S^{0.54}$	$\left(\frac{3.59Q}{C_hd^{2.63}}\right)^{1.852}$

Coefficients

Character of Pipe	Hazen-Williams coefficients of roughness, C_h	
	New Installations	Design Values
PVC, Fiberglass, polyethylene, pipe with glass lining	150	140-150
Ductile iron, steel and concrete pipe with centrifugally spun concrete or bituminous lining, asbestos cement pipe	140	135-140
Pipe cement lined in place with good workmanship, 24-inch and larger	130	120-130
Pipe cement mortar lined in place, ordinary workmanship, smaller than 24-inch	120	115-125
Existing unlined cast iron pipe[1]		80-120

[1]Design C factors vary, and may require determination by field measurement.

3.2 Comparison of Formulas Commonly Used

Pipeline systems for water should be designed for maximum flows with reasonable velocities, or the friction losses will be extreme. Under these limitations of velocity, the various formulas are in general agreement. Capacities of commercial pipe can vary from the theoretical by several percent, depending upon actual internal diameter and the lining process. In addition, all types of metering devices have some inherent error. As a result, a good hydraulic analysis of the piping system with proper application of a roughness coefficient throughout may be but 95% correct. Within the velocity limits specified, the various standard formulas with their respective roughness coefficients will be in the same range of accuracy; the following comparisons for 12-in. pipe and 24-in. pipe summarize the dispersion of results:

12-in. Diameter Pipe v=5fps				Loss Per 1,000 ft.
Hazen-Williams	C_h	-	135	6.79 ft
Darcy-Weisbach	ϵ	-	0.0004 ft	6.77 "
Manning	n	-	0.010	7.18 "
Scobey	C_s	-	0.400	7.00 "

24-in. Diameter Pipe V=5fps				
Hazen-Williams	C_h	-	135	3.03 ft
Darcy-Weisbach	ϵ	-	0.0004 ft	2.92 "
Manning	n	-	0.010	2.85 "
Scobey	C_s	-	0.400	2.94 "

The foregoing shows that the calculated energy or friction losses vary approximately 3% from the average, or a maximum difference of 6% for a velocity of 5 fps in the range of diameters investigated. The differences in the results are greater at higher velocities (or higher magnitudes of flow) but reasonably so in the range of rational friction losses. The difference at lower velocities are usually not as important for those magnitudes of flow which result in low energy losses.

3.3 Water Pipe Interiors

The interior of a pipeline, transmitting water, can be changed after installation by actions relating to the physical and chemical characteristics of the water. The result of these actions may reduce the actual diameter of the opening and alter the original smoothness of the interior of the pipe, both of which would reduce the capacity of the pipe.

Conditions which would adversely affect the flow in a pipe after it is placed in operation are: sedimentation, organic growths, mineral deposits, and corrosion.

Sedimentation may occur when a raw water supply contains sand or silt which settle in the pipe when the velocity of flow is reduced. This condition may be remedied by flushing and cleaning the pipeline and is not related to pipeline hydraulic design.

Some waters, from both surface and ground supplies may cause the growth of organisms or bacterial slimes on the interior of the pipe, which will reduce the rate of flow in the line. This is an operational problem and can be corrected by the proper use of chemicals such as chlorine in both type and amount. This type of growth is not related to the hydraulic design of the pipeline.

Corrosion can occur in the interior of an unlined metallic water main due to an active corrosive water. Current practice is to line metallic pipelines to inhibit corrosion. Cement mortar is the most common lining material used to maintain the initial pipe capacity. Other lining materials are also available. Coal tar is not recommended for lining potable water pipes.

Mineral deposits may coat the interior of a pipeline carrying water with a high mineral content. For water supply use it is usually necessary to remove such minerals by appropriate chemical treatment. To avoid loss in pipe capacity, the treatment should be located ahead of the transmission pipeline.

To determine the hydraulic capacity of old pipelines, flow tests on representative sections of the pipeline will disclose the approximate roughness conditions of the pipe interiors. Procedures for such tests are described in some textbooks on water supply, and are based on a measured flow through a straight section of pipeline (as long as possible) while recording the pressure drop by gages installed at both ends of the section. Information as to interior wall conditions and diameter can be noted from an inspection of any section of the pipe which has been removed for any reason or from measurements made through special corporation taps using inside diameter calipers.

3.4 Minor Losses

Water flowing in a straight pipe under pressure at a constant velocity produces a constant friction loss. A change in the velocity or direction of flow creates additional energy losses. Such losses are classified as minor, as they are a small percentage of the total energy loss in relatively long pipelines. A pipeline with a length in excess of 500 diameters is usually classified as "long." Minor losses become important when small diameter pipes and high velocities are considered in the total piping and appurtenances relating to a pumping station. Minor losses are produced by flow through bends, fittings, valves and openings such as fire hydrants and in and out of reservoirs.

The loss of head h_m (ft or m) due to change in pipe alignment or in an appurtenance can be expressed as follows:

$$h_m = K \frac{V^2}{2g} \qquad (3)$$

in which K is the dimensionless minor loss coefficient through the pipeline fitting or appurtenance and V, as previously defined, is the mean velocity in the pipe under consideration. The expression $V^2/2g$ is termed the velocity head. Many studies and tests have been conducted to determine the approximate value of minor loss coefficients for various fittings and appurtenances.

The equivalent length of pipe is the number of feet of straight pipe which will cause the same pressure drop as the flow through the fitting or appurtenance. The relationship between pipe diameter (d, D) and pipe length (l) within the hydraulic formulas allows the minor losses to be stated in terms of the equivalent length in pipe diameters rather than the minor loss coefficient (K). Examples of the value of K and equivalent lengths in pipe diameters for various fittings and appurtenances are listed in Table 5. Actual losses may vary.

TABLE 5 - LOSSES IN PIPE FITTINGS AND APPURTENANCES

Description of Pipe Fitting or Appurtenance	Minor Loss Coefficient (K)	Loss in Equivalent Length in Pipe Diameters (d)
Gate Valve		
3/4 Closed	24.0	900
1/2 Close	5.6	160
1/4 Closed	1.2	35
Full Open	0.2	13
Angle Valve Open	2.5	170
Globe Valve Open	10.0	340
Swing Check Valve	0.6 - 2.5	80
Elbows		
90° Standard	0.7	32
90° Long Radius	0.6	20
45° Standard	0.5	16
Tee Flow Through Run	1.8	20
Sudden Contraction*		
d/D = 1/4	0.4	15
= 1/2	0.3	12
= 3/4	0.2	7
Sudden Enlargement*		
d/D = 1/4	0.9	32
= 1/2	0.6	20
= 3/4	0.2	7
Flow From Storage Tank		
Pipe Projecting Into Tank	0.78	
Pipe Flush With Tank Wall	0.5	
Slightly Rounded Entrance	0.23	
Well Rounded Entrance	0.04	
Flow Into Storage Tank		
All Conditions	1.0	

* For Contractions and Enlargements, coefficients are applied to the conditions in the smaller diameter pipe.

The waterworks industry now makes use of various types of valves for sectionalizing or isolating portions of main lines or for control purposes. Each has distinctive interior contours which has an effect on fluid flow. Flow rate and loss of head charts have been prepared by the various manufacturers for all sizes and types of valves.

Water meters in common usage in the waterworks field for small diameter pipelines are positive displacement type, turbine type and velocity or current meter type or a combination of two of the three for compound meters. Larger diameter pipelines and related larger flows can be monitored by pressure differential meters such as a Venturi tube, orifice assembly or magnetic flow meters which monitor the flow with no obstruction within the tube. Pressure differential or magnetic types of flow meter installations cause small friction losses. However, small diameter meters with internal moving parts may be the source of considerable amounts of head loss in the piping system, particularly if the meter size has been reduced from the supply main size.

Attention should be directed to meter selection, particularly in the design of pumping stations. The meter manufacturer should be requested to furnish certified curves as to the performance of the proposed meter so that the total friction loss calculations can be completed.

REFERENCES

1. King, H.W., Wisler, C.O., Woodburn, J.G., "General Method of Determining f," Hydraulics, 5th ed., John Wiley & Sons, Inc., New York, 1958, pp. 189-194.

2. Ibid, pp. 224-227.

3. Whitaker, S., "Friction Factors: Experimental," Introduction to Fluid Mechanics, Prentice-Hall, Inc., Englewood Cliffs, N.J., 1968, pp. 293-306.

4. Williams, G.S. and Hazen, A., Hydraulic Tables, John Wiley & Sons, Inc., New York, 1945.

5. Hydraulic Institute, Hydraulic Institute Standards. 13th ed., Cleveland, Ohio.

6. King, H.W., Brater, E.F., "Handbook of Hydraulics," 5th ed., McGraw-Hill Book Company.

SECTION 4 - PUMPED PIPELINES AND SURGE CONTROL

The large majority of pipelines and water distribution systems incorporate pumps to move water through the system and maintain required pressures. For this reason, this section will address several aspects of pumps and pumped pipelines. Further, because a great many surge or water hammer problems are associated with pumped pipelines, that topic will be covered as well. The information provided will serve the user directly in the design of pipelines and also provide other sources of material for investigating potential problems in greater depth.

4.1 Pumps in Pipe Systems

Pumps in a pipeline commonly occur at the upstream end of a pipeline adjacent to a well or reservoir (source pump) and at intermediate locations in the system (booster pumps). In each case, the discharge through the pump and the head increase across the pump are affected by the pipe system in which the pump is installed.

4.1.1 Types of Pumps

Pumps are divided into two broad categories -- dynamic pumps and displacement pumps (Karassik, Krutzsch, Fraser, and Messina [7]. Displacement pumps consist of reciprocating pumps (piston, diaphragm actuated) and rotary pumps (gear, lobe, screw, etc. actuated). Displacement pumps are seldom used in water supply pipelines so they will not be discussed further.

Dynamic pumps include centrifugal pumps (radial, mixed and axial flow) and specialty pumps (jet pumps, air lift and hydraulic ram).

The centrifugal type of pump develops head utilizing a rotating impeller (or propeller) within a confined circular casing. The impeller (or propeller) is comprised of a number of vanes which force the liquid to the outside of the casing by centrifugal action or normal to the plane of the blades by a propeller-type action. Some pumps combine both such actions through vane curvature and orientation.

A single-stage pump has only one impeller; a multistage has two or more impellers in a single casing. The three general classes of centrifugal pumps are radial flow, mixed-flow, and axial flow. Descriptions of each are given in Karassik, et al [7] and Sanks, et al [8]. Water supply pipelines and distribution systems almost exclusively use centrifugal-type pumps, so only this classification of pumps will be addressed.

4.1.2 Pumps in a Pipeline

To analyze pumped pipelines, the work-energy equation must be modified to include the energy added by the pump.

$$z_1 + \frac{p_1}{\gamma} + \frac{V_1^2}{2g} + H_p = z_2 + \frac{p_2}{\gamma} + \frac{V_2^2}{2g} + \sum h_{L_{1-2}} \quad . \quad . \quad (4.1)$$

where H_p is the energy added by the pump to each pound of liquid passing through the pump. In the case of multistaged or series pumps, H_p is the sum of the head increases across each pump or stage.

The head increase across the pump is a function of discharge through the pump and is determined experimentally by the pump manufacturer. The information is presented graphically on a diagram known as the characteristic diagram. Information on power requirements and pump efficiency at varying discharges is generally included as well. Often data for more than one size of similar pumps are included. Figure 1 shows a typical characteristic diagram from a commercially available pump.

Because power requirements for a pumping situation are of interest, it is important to know the relationship between energy added to the water (H_p) and power supplied to the pump. The power added to the water as it passes through the pump is

$$WHP = \frac{Q\gamma H_p}{550} \quad . \quad . \quad . \quad . \quad . \quad . \quad . \quad (4.2)$$

where WHP is the horsepower added to the water. The power supplied to the pump shaft differs from the above because of friction and leakage losses but is related through the pump efficiency

$$WHP = \eta BHP \quad . \quad . \quad . \quad . \quad . \quad . \quad . \quad (4.3)$$

where BHP is the horsepower applied to the pump shaft. The overall pump efficiency is in the range of 70% to 90% and is found from laboratory tests for a number of different operating conditions.

Two examples are provided to demonstrate how the

pump curve graphical data can be used in conjunction with the work-energy equation to determine the flow rate in a system. While a graphical solution to the problem (addressed next) is the more common approach, the numerical solution is provided to clearly show how parallel and multi-staged pumps are handled.

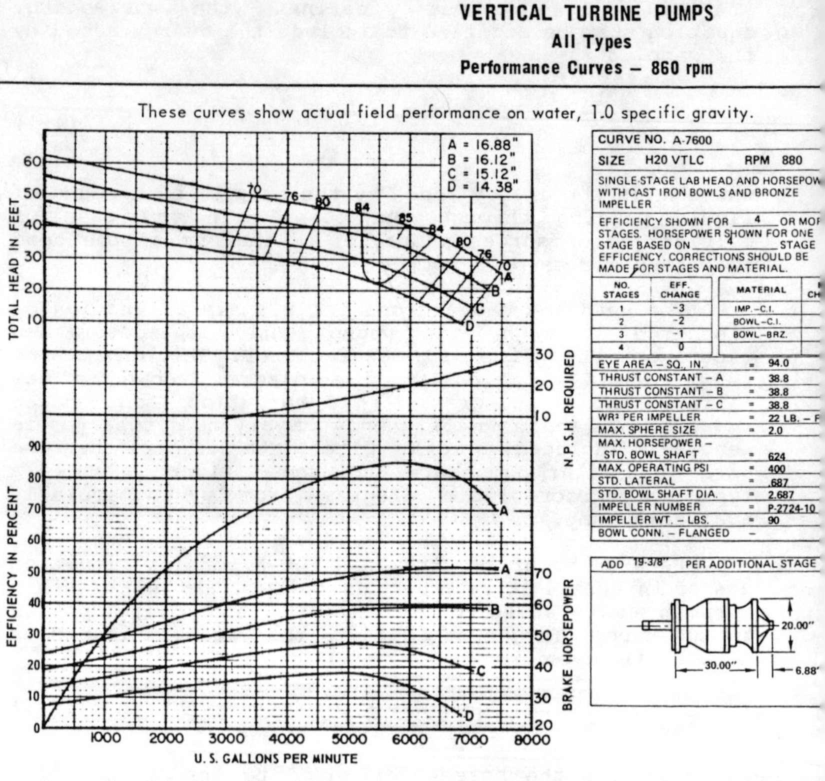

Figure 1
Typical Pump Characteristic Diagram. (Courtesy of Allis-Chalmers)

Example 4.1

A single-stage pump with the characteristics of Figure 1 (Curve A) is to be used as a source pump to move water from one reservoir at elevation 210 feet to a canal at elevation 245 feet. The line is 14,250 feet long and has an inside diameter of 35.39 inches. The Darcy f-value is 0.017.

Neglecting minor losses in the pipeline and pump station, compute the discharge in the pipeline, the pump efficiency and the horsepower requirements for the prime mover.

Solution

From Equation 4.1

$$z_1 + \frac{p_1}{\gamma} + \frac{V_1^2}{2g} + H_p = z_2 + \frac{p_2}{\gamma} + \frac{V_2^2}{2g} + \Sigma h_{L_{1-2}}$$

$$210 + 0 + 0 + H_p = 245 + 0 + 0 + f\frac{L}{D}\frac{V^2}{2g}$$

$$H_p = (245 - 210) + \frac{.017 \times 14{,}250}{\frac{35.39}{12}} \frac{V^2}{64.4}$$

$$H_p = 35 + 1.28V^2$$

This equation must be solved by trial in conjunction with Figure 1. The best approach is to use a trial solution table.

Assume Q(gpm)	V (fps)	$35 + 1.28V^2$	H_p(ft)
5000	1.63	38.4	45
7000	2.28	41.7	32
6000	1.96	39.9	40

The solution is Q = 6000 gpm.

From Figure 1, the efficiency is 85 - 3 = 82% and a motor of 72 horsepower will be required. Note from Figure 1, it is necessary to adjust efficiency for a pump with less than four stages.

Example 4.2

Solve the problem of example 4.1 using two three-stage parallel pumps with characteristics given by Curve D.

Solution

The hydraulic equation for the pipeline would be the same, but the solution in conjunction with the pump characteristic must take the multistaging and paralleling effect into account

$$H_p = 35 + 1.28V^2$$

Assume Pump Q (gpm)	Qline (gpm)	V(fps)	$35 + 1.28V^2$	H_p/stage	H_p(ft)
4000	8000	2.61	43.7	29	87
6000	12,000	3.91	54.6	18	54

The solution is Q = 12,000 gpm.

From Figure 1 the efficiency is 85 - 1 = 84% and a motor of 34 x 3 = 102 horsepower will be required for each pump.

4.1.3 System Head Requirement Curves

Often the engineer is required to consider a variety of operating conditions for the pipeline and determine how those conditions will affect the head and discharge (and by implication, power requirements and efficiency) of the pumps. A graphical method for accomplishing this is particularly attractive, especially if the number of pumps operating varies and if the water levels of the reservoirs or the pressure requirements fluctuate.

The procedure is to write the work-energy equation 4.1 for each of the anticipated operating conditions and plot the resulting H_p vs. Q data. These are known as system demand curves. Next, the pump characteristics are expressed in terms of total head increase and total line flow rate and a curve is plotted for each operating situation. These are known as pump supply curves. The points where the system demand and pump supply curves intersect are operating conditions for that particular configuration.

The procedure is best illustrated by graphically

solving Example 4.2.

Example 4.3

Solve Example 4.2 for one or two pumps operating and for supply reservoirs at the high water mark of 210 feet and the low water mark of 200 feet.

Solution

The system demand curves are prepared from the work-energy equation and are as follows:

High water - $H_p = 35 + 1.36 \times 10^{-7} Q^2$ (gpm)
Low water - $H_p = 45 + 1.36 \times 10^{-7} Q^2$ (gpm)

Discharge data from the pump curve would have to be multiplied by 2 (2 pumps in parallel) and head data by 3 (3 stage) on order to plot as H_p vs. Q.

The resulting curves are shown on Figure 2.

The solution to Example 4.2 is shown at the Q = 12,000 gpm point. The curves show that if the water level is at the low point in the supply reservoir, the pumps will only produce 11,400 gpm.

Under the worst conditions with one pump running at low reservoir level, only 6200 gpm is pumped.

4.1.4 Submergence and Cavitation Considerations

Consideration must be given to the suction conditions for the pump installation. Care must be taken to place the inlet to the pump sufficiently far below the water surface of the sump or reservoir to prevent air being sucked into the inlet. This problem generally manifests itself through vortexes forming near the pump inlet. The Hydraulics Institute [6] publishes general guidelines for the required submergence of pump inlets. Most pump manufacturers also provide data which sets forth submergence requirements for their line of pumps. Figure 3 is an example of what may be provided by the manufacturer. For example, considering the pump of Example 4.1, Figure 3 gives a required submergence of 4.5 feet to prevent vortexing at the maximum discharge of 7500 gpm.

Cavitation in pumps must be prevented at all costs for the complete range of anticipated pump discharges. This is generally accomplished by specifying a submergence value based on the manufacturer's recommendations. While the pump vendor is quite capable of assisting in this process, it is good practice for the engineer to make an independent evaluation.

The occurrence of cavitation is determined in the testing laboratory. When the discharge and pump location are such that the pressure at some point in the pump drops to the vapor pressure of the liquid, cavitation

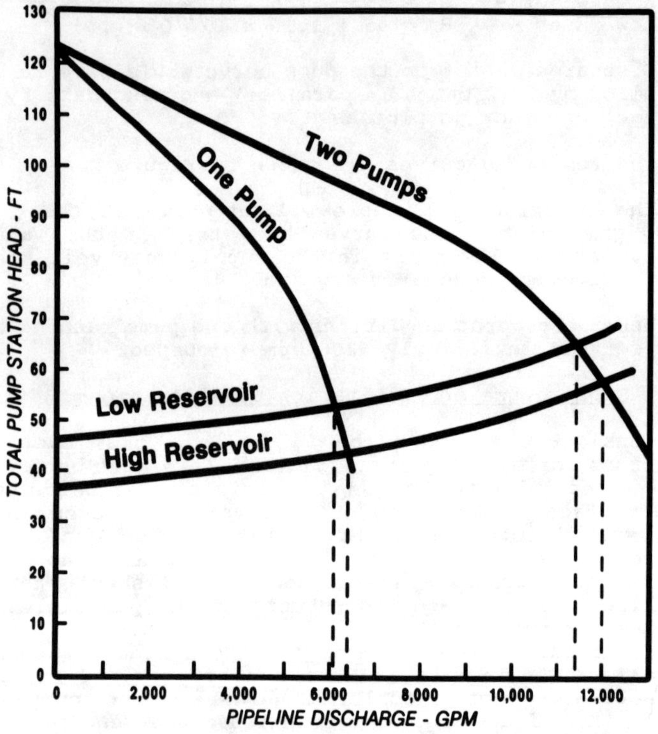

Figure 2
Pump Supply and System Demand Curves

VERTICAL TURBINE PUMP SUBMERGENCE

TO PREVENT VORTEXING

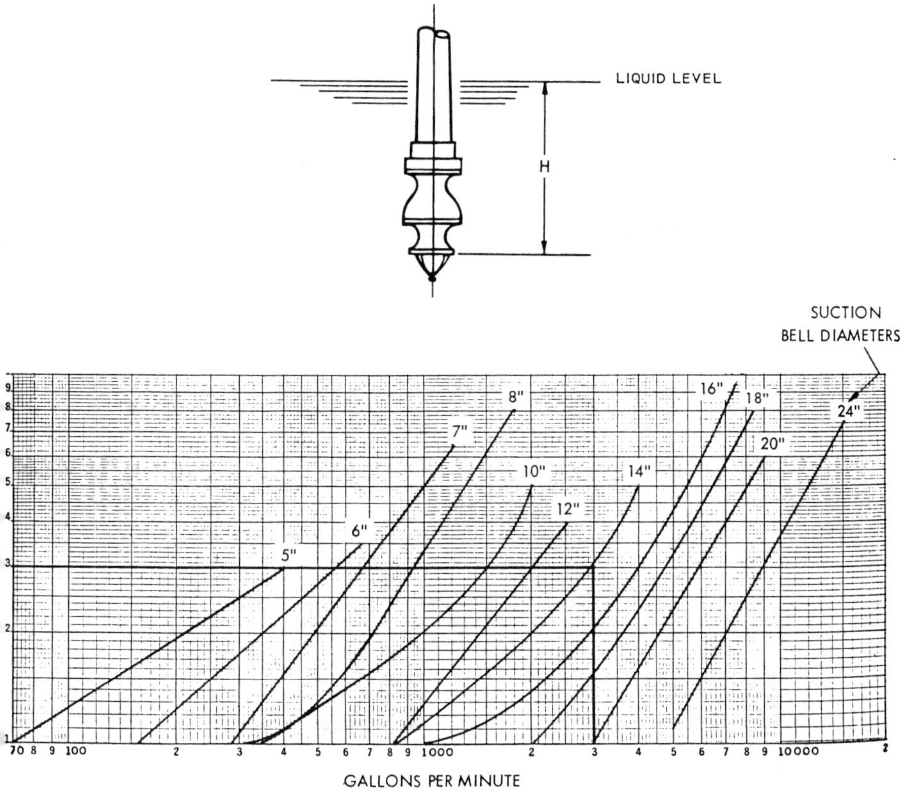

EXAMPLE: 14 in. pump — 3000 gpm requires a minimum of 3 feet of submergence.

NOTE: This chart does not cover NPSH requirements; check curve in each case. NPSH may require additional submergence.

Figure 3
Submergence Values for Turbine Pumps (Courtesy of Allis-Chalmers)

begins. A further deterioration in the conditions will lead to massive cavitation and result in sharp drops in efficiency, possible vibration problems and, if continued for long periods, failure of the pump hardware.

Information which can be used to design pump installations to prevent degradation of performance as the result of cavitation is given by pump manufacturers in several ways. The simplest and least useful is "suction lift." This value is supposed to be the maximum height the pump may be situated above the sump to meet the manufacturer's guaranteed performance curve. If one has unusual conditions, it is not clear what to do.

The most fundamental measure of the conditions necessary to prevent cavitation-related performance degradation is the required net positive suction head $NPSH_R$. An equation for $NPSH_R$ is

$$NPSH_R = \frac{p_o}{\gamma} - \frac{p_v}{\gamma} - z_s - h_{L_{ent}} \quad . \; . \; . \; . \; . \; . \; . \; . \quad (4.4)$$

where p_o is the absolute ambient pressure (generally atmospheric), p_v is the vapor pressure of the liquid, z_s is the elevation of the impeller above the supply water surface, and $h_{L_{ent}}$ is the head loss between the sump and the pump impeller. A curve of $NPSH_R$ versus discharge is given by some manufacturers (see Figure 1). With a value of $NPSH_R$, Equation 4.4 can be used to provide information on z_s for nonstandard values of p_o and p_v.

It should be noted that cavitation and cavitation damage can occur in a pump even though it is operating within the manufacturer's $NPSH_R$ requirements. This occurs because cavitation will begin before performance degradation becomes measurable. If operation occurs at discharges well above or below the normal capacity, prevention of cavitation may require NPSH values several feet larger than indicated by the manufacturer. Under these conditions, consideration should be given to assigning conservative values to submergence.

Example 4.4

For the pump of Example 4.1, determine how far the lowest impeller must be placed below the water supply surface for the highest indicated discharge. Water at $60^{\circ}F$ is being pumped from a reservoir at elevation 5000 feet above sea level.

Solution

From Figure 1, at a discharge of 7500 gpm, the value of NPSH required is 28 feet. From any fluid mechanics or hydraulics reference, one can determine that p_v = 0.26 psia and p_o = 12.2 psia. Assuming that h_{Lent} is included in the $NPSH_R$ value, one can compute z_s by substituting in Equation 4.4

$$28 = 12.2 \times 2.31 - 0.26 \times 2.31 - z_s$$

$$z_s = 28.2 - 28.0 - 0.6$$

$$z_s = -0.4 \text{ ft.}$$

To meet the manufacturer's $NPSH_R$ requirements, the impeller need only be 0.4 feet below the water surface. In this case, however, air vortexing would control to require a submergence of 4.5 feet, based on Fig. 3.

4.1.5 Pump Similarity Relationships

The principles of hydraulic similitude find important application in both constant-speed pumping situations and in water hammer situations caused by pump power failure. The similarity principles permit us to predict heads, discharges, and torques over a wide range of speeds when using only the basic information given at one speed.

For those familiar with conventional modeling parameters for hydraulic problems, it is interesting to recognize that in the case of pumps, kinematic similarity takes precedence over dynamic similarity. Flow velocities are proportional to Q/D^2 and impeller velocities are proportional to ND, where D is a typical pump dimension (generally the impeller diameter) and N is the rotational speed in revolutions per minute (rpm). Kinematic similarity requires these ratios to be constant, which gives the following equations:

$$\frac{Q}{ND^3} = \text{Constant} \quad . \; . \; . \; . \; . \; . \quad (4.5a)$$

or

$$\frac{Q_1}{N_1 D_1^{\,3}} = \frac{Q_2}{N_2 D_2^{\,3}} \quad . \; . \; . \; . \; . \; . \quad (4.5b)$$

Additional equations relating head, torque and power to speed and size are as follows

$$\frac{H_{p_1}}{N_1^2 D_1^2} = \frac{H_{p_2}}{N_2^2 D_2^2} \quad . \; . \; . \; . \; . \; . \; . \; . \quad (4.6)$$

$$\frac{T_1}{N_1^2 D_1^5} = \frac{T_2}{N_2^2 D_2^5} \quad . \; . \; . \; . \; . \; . \; . \; . \quad (4.7)$$

$$\frac{BHP_1}{N_1^3 D_1^5} = \frac{BHP_2}{N_2^3 D_2^5} \quad . \; . \; . \; . \; . \; . \; . \; . \quad (4.8)$$

These relationships can be used to predict changes in discharge, head, and power requirements for pumps of different size (impellers trimmed) or pumps running at different speeds. The accuracy of the prediction decreases somewhat as the changes in speed and size range further from the test conditions. The application we are most interested in for water hammer situations is when power to a pump fails and the pump begins to decelerate. In this case, the decelerating pump discharge, head and torque can be computed from Equations 4.5, 4.6 and 4.7 using D_1/D_2 = 1.0 (same machine) at successively changing speeds. Under these conditions we use the similarity rules as the best estimate of pump performance at any speed.

4.1.6 Pump Selection

In the initial stages of pump selection, it is often helpful to get some idea of the type of pump or the configuration of the pumping installation that will be required. This is particularly true if there are going to be several pumps in the installation. A traditional means for accomplishing this employs the concept of specific speed.

Specific speed N_s is defined as follows:

$$N_s = \frac{N\sqrt{Q}}{H^{3/4}} \quad . \; . \; . \; . \; . \; . \; . \; . \quad (4.9)$$

where N = pump speed in revolutions per minute, Q = pump discharge in gallons per minute, and H = pump head increase in feet. Equation 4.9 is for single stage, single suction pumps. For double suction pumps, use one-half the pump flow rate. For multistage pumps, use the head per stage.

Specific speed values are computed for pumps at their points of maximum efficiency and the numerical value is typical of the type of pump. For example, pumps with N_s - values less than 1000 are generally radial-flow centrifugal pumps. Those with values greater than 10,000 are generally axial flow (propeller) pumps. Knowing the specific speed of a pump permits one to estimate its efficiency and to classify it as to type. Figure 4 illustrates the relationship between specific speed and pump efficiency and type.

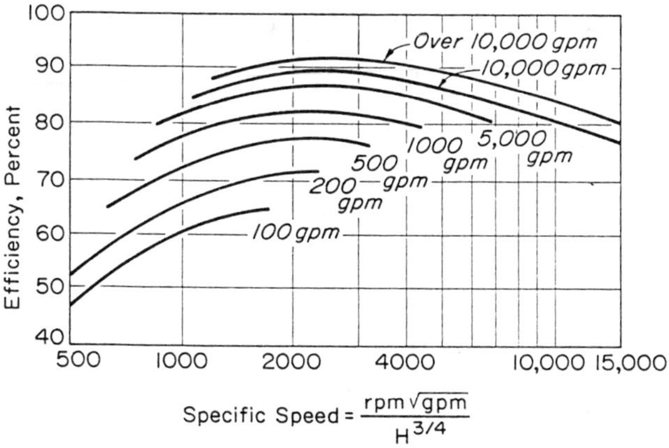

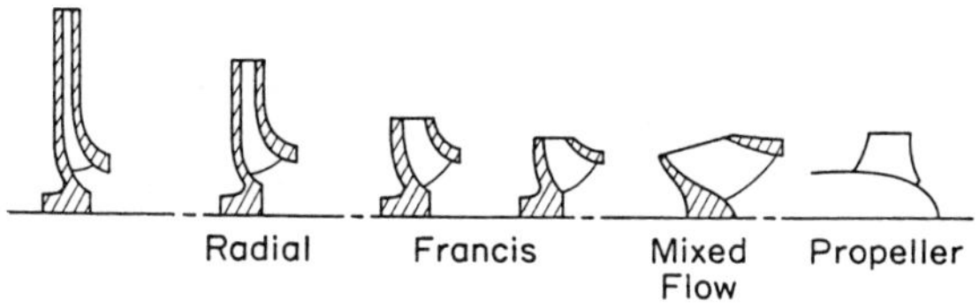

Figure 4
Efficiency and Impeller Type as a Function of Specific Speed

The value of the specific speed concept is in initially deciding on a pumping configuration. One computes the specific speed for a situation, then uses that value to select the number and type of pumps. The following example illustrates the application.

Example 4.5

An irrigation project requires that 11,000 gpm be pumped from a river at elevation 395 ft to a reservoir at elevation 840 ft. The line length is 28,400 ft. Select a preliminary pumping configuration.

The pumps will be required to provide enough head to meet the static lift of 445 ft plus the friction loss in the line. Estimating a velocity of 5 fps and a Darcy f-value of .018, the friction loss is 80 ft.

Using a typical pump speed of 1760 rpm, the specific speed is calculated as

$$\text{(Eq. 4.9)} \quad N_s = \frac{1760\sqrt{11{,}000}}{525^{3/4}} = 1683$$

From Figure 4, a single pump of the Francis or turbine type would work but it would be difficult to find one of this capacity and head capability.

Assume in this case that redundancy and flexibility require a minimum of four pumps. Under these conditions,

$$N_s = \frac{1760\sqrt{2750}}{525^{3/4}} = 841$$

This value of specific speed calls for a radial-flow centrifugal pump, however, this is not a specific speed which gives high efficiency. So, multistaging can be used to increase the specific speed. For a flow rate of 2750 gpm, a specific speed of about 2500 gives the best efficiency (Fig. 4).

This time Eq 4.9 is solved for the head per stage H.

$$2500 = \frac{1760\sqrt{2750}}{H^{3/4}}$$

$$H = 123 \text{ ft per stage}$$

$$\text{Number of stages} = \frac{525}{123} = 4.3 \text{ (or 5 stages)}$$

It appears the best pumping configuration will be four 5-stage turbine pumps. The next move is to look in the pump catalogs for a pump running at 1760 rpm which will produce 123 ft per stage at 2750 gpm at peak efficiency.

It is clear from the above example, that the specific speed can be made to vary over a wide range, depending on such arbitrary conditions as motor speed and redundancy in system design. Hence, the use of specific speed as a design tool is primarily a rational means of arriving at a rough approximation of a workable pumping configuration.

4.1.7 Pumps in Wastewater Systems

While wastewaters are generally very similar to water in hydraulic characteristics, other physical and chemical characteristics may require special considerations in choosing wastewater pumping equipment. Raw sanitary sewage and some industrial wastes may contain gross solids, limiting the minimum size and type of pumping equipment that may be used. Industrial wastewaters may be mildly to severely corrosive, limiting pumping equipment to hardware made of corrosion resistant materials.

Centrifugal pumps used on wastewaters containing solids are generally of the single suction volute type with overhung impeller. Radial non-clog design units are used in sizes up to 10 in., with capacities of 40 gpm to 6000 gpm and heads of 40 ft to 100 ft. Mixed flow design units are available in sizes from 8 in. to 84 in. with capacities of 1000 to 80,000 gpm and heads of 10 ft to 60 ft.

Centrifugal pumps under 4 in. in size are generally not used since they may clog very readily. Pumps from 4 in. to 10 in. in size are generally of the "non-clog" design; casings and impellers are designed to permit passage of solids, stringy material and grease. The casing throats and impellers are wide, with the impeller having only one or two vanes. In some pumps the impeller may even be recessed in the casing, imparting centrifugal force to the liquid without being immersed in it. Compared to pumps for clear water, these "non-clog" pumps sacrifice efficiency for the ability to pump solids without clogging.

4.1.8 Pump Stations

The design of pump stations is, by itself, the subject for an entire design manual. Traditionally, the Hydraulic Institute Standards [6] have been used as the guidelines for pump intake design. The configuration of the pump station itself, the electrical works, the valving, the control systems and the operation procedures are spelled out in considerable detail in Pumping Station Design by Sanks, et al [8]. The reader is referred to that thorough work to accomplish the station design.

4.2 Hydraulic Transients

Unsteady flow problems in engineering practice are of significant importance because they can cause excessive pressure, noise, and material damage far beyond that indicated by steady flow analysis. In fact, the problems created by hydraulic transients may be so severe as to constitute actual or performance failure of a system.

In this work hydraulic transient considerations will be confined to variations in pressures and velocities in piping systems. The transient condition may result from a change in valve setting, air exhaustion, power failure to a pump, etc. In a general sense, any action resulting in a change in velocity in the system will create pressure waves which will propagate throughout the system. The more abrupt and severe the velocity change, the more dramatic the pressure change.

The analysis of unsteady flow in piping systems can be divided into two general approaches. The first, called rigid water column theory, considers the fluid and pipe as inelastic with instantaneous propagation of pressure changes throughout the system. The resulting differential equations are "ordinary" and relatively easy to solve. The second approach, called elastic theory, treats the fluid and the pipe as elastic materials and is commonly referred to as "water hammer." In water hammer the pressure increments resulting from velocity changes depend on the elastic properties of the pipe and the fluid and they propagate throughout the system in a manner determined by these elastic properties. While more realistically representing the true transient phenomena, the elastic theory leads to a pair of partial differential equations which are most commonly solved by numerical techniques such as the method of characteristics.

In order to decide which technique of analysis to employ, it is necessary to learn under what conditions the simpler rigid water column theory can be usefully employed. Remember, water is not incompressible, nor are pipe walls inelastic, hence rigid water column theory is always an approximation.

To set the stage for further understanding, consider what occurs when both the water and the pipe are allowed to behave in a natural manner. To further simplify, ignore friction and work with a horizontal, constant diameter pipe. This will be called "simple water hammer."

4.2.1 Simple Water Hammer

When the pressure pulses generated by a change in velocity in a pipe system propagate throughout the system, "water hammer" is said to be occurring. Sometimes the pressure pulses are severe enough to cause pipes to strike against one another, the walls, or pipe hangers, or to cause slack valve fittings to lash. The noise resulting from this action undoubtedly is responsible for the term "water hammer."

The magnitude and duration of the pressure surge created depend on the magnitude of the change in velocity, the time interval over which the velocity changes, and the physical properties of the fluid and pipeline system in which the change is occurring.

To get an elementary understanding of the phenomenon consider the system shown in Figure 5. Note the hydraulic grade line (HGL) and the energy line (EL) are considered coincident because velocity head will be quite small in comparison to water hammer pressure changes. They will be combined and referred to as the EL-HGL. The head **H** at the reservoir causes a steady velocity of **V** to occur in the long pipe downstream. Only occurrences in the first section of the pipe **L** feet long between the reservoir and the valve will be examined.

In the system shown in Figure 5 the valve is closed suddenly and the resulting increase in pressure head ΔH propagates upstream at the wave speed **a**. For the time being, what happens downstream of the valve will be ignored. The time required for the wave to travel to the upstream reservoir is **L/a** seconds. The sequence of events occurring after the valve closes is shown in Figure 5 for increasing time in terms of **L/a**.

As the wave moves back and forth throughout the system, the pressure head at any section of the pipe fluctuates between $H \pm \Delta H$. Figure 6 shows how this pressure head varies with time at three sections of the pipe.

Two important ideas can be developed from Figure 6. First, **L/a** is the important time parameter in the system. Second, the pressure at a point remains at its new value (after valve closure) until (1) further changes in velocity create new pressure waves which propagate to the point, or (2) the original pressure wave reflects back to the point from a reservoir or some other change in the piping system. Note that if the valve were closed in any time less than 2 **L/a**(t_c), the full water hammer pressure would develop at the valve. It is clear that very high pressures can be created without suddenly closing valves.

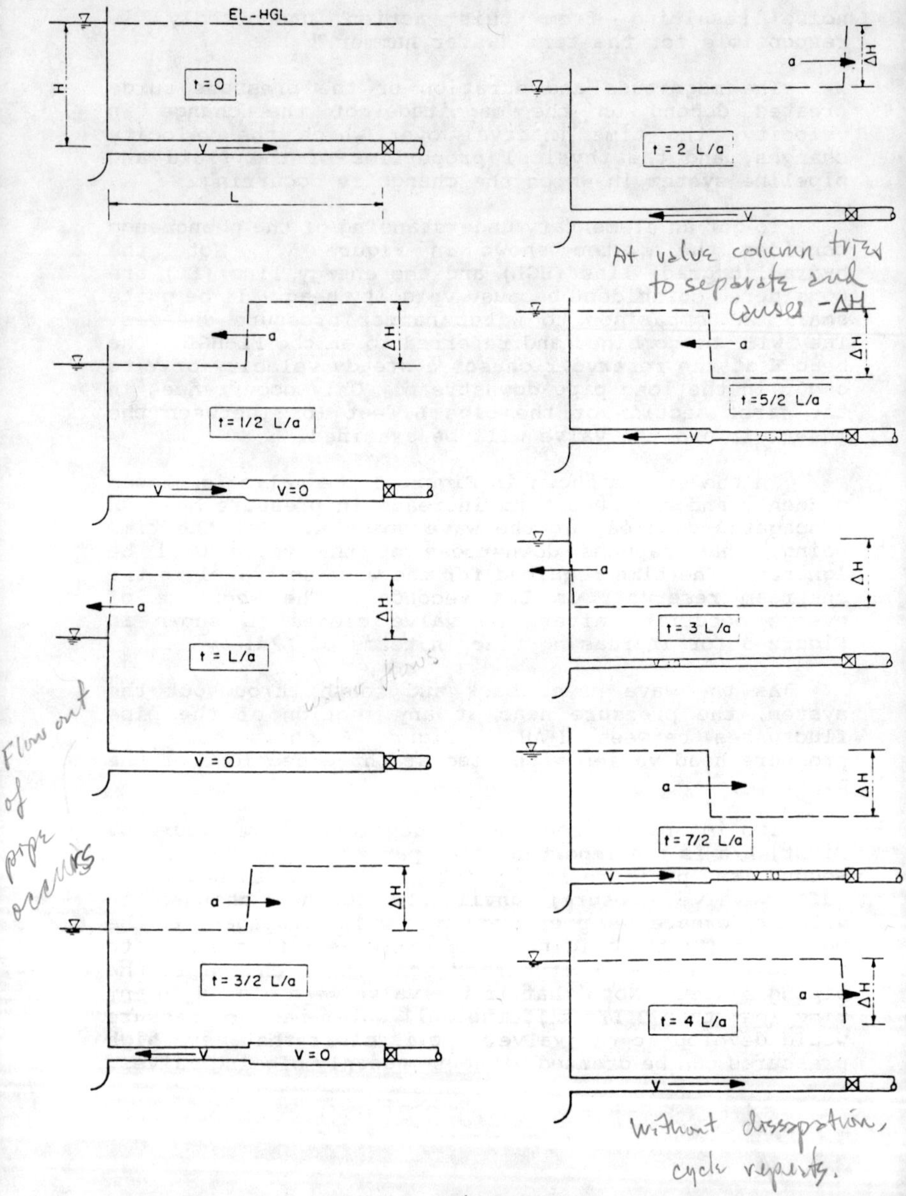

Figure 5.
Pressure waves in a simple pipe resulting from sudden valve closure. (Watters [9])

4.2.2 Rigid Water Column Theory

More commonly known as unsteady flow, rigid water column theory is developed from Newton's second law F = ma as applied to a column (i.e., a pipe-full) of water.

The equation resulting from this application of Newton's second law to the simple pipe system of Figure 7 is

$$\frac{p_1}{\gamma} - \frac{p_2}{\gamma} - \frac{fL}{2gD}V^2 = \frac{L}{g}\frac{dV}{dt} \quad \ldots\ldots\ldots \quad (4.10)$$

where
- p = pressure
- f = Darcy-Weisbach friction factor
- L = pipe length
- γ = specific weight of fluid
- D = pipe diameter
- V = flow velocity
- g = acceleration of gravity
- dV/dt = fluid acceleration

Flow Establishment

As an application of rigid water column theory, consider flow establishment in the pipe of Figure 7. Initially the fluid is at rest in the pipe with a pressure head of H_o throughout. The valve is instantly opened resulting in atmospheric pressure at point 2. Equation 4.10 modified to this application is

$$H_0 - \frac{fL}{2gD}V^2 = \frac{L}{g}\frac{dV}{dt} \quad \ldots\ldots\ldots \quad (4.11)$$

Theoretically, the time required to reach steady state velocity V_0 is infinite. However, integration of the equation shows the time required to reach 99 percent of the steady state discharge is

$$t_{99} = 2.65\frac{LV_0}{gH_0} \quad \ldots\ldots\ldots \quad (4.12)$$

Pressure Caused by Valve Closure

If the steady state flow just established is to be shut off by closing the valve, Equation 4.10 gives

$$H_0 - \frac{p_2}{\gamma} - \frac{fL}{2gD}V^2 = \frac{L}{g}\frac{dV}{dt} \quad \ldots\ldots\ldots \quad (4.13)$$

where

$$H_o - \left(K_L(t) + \frac{fL}{D}\right)\frac{V^2}{2g} = \frac{L}{g}\cdot\frac{dV}{dt}$$

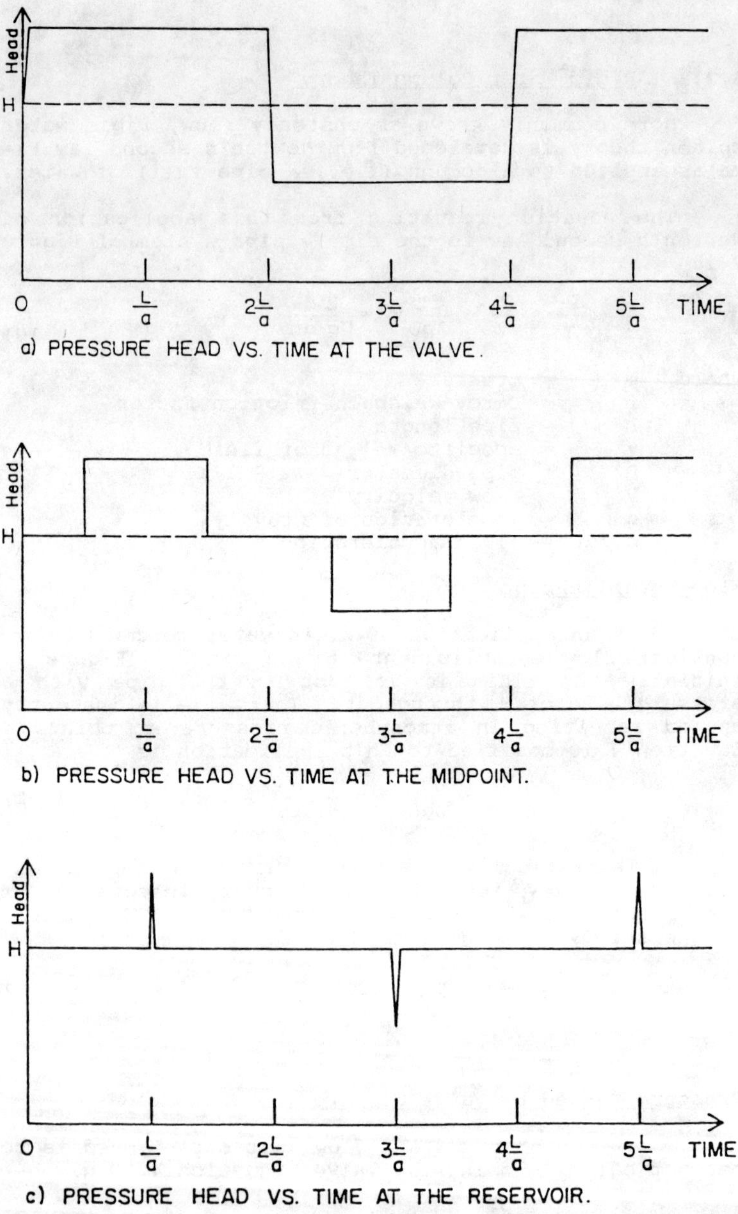

Figure 6.
Pressure head vs. time at points along the pipe (Watters [9]).

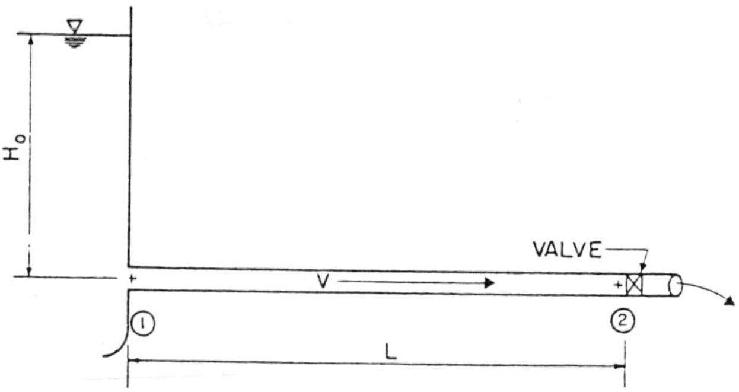

Figure 7.
Simple system for applying rigid water column theory (Watters [9]).

$K_L = f(t)$

$$\frac{p_2}{\gamma} = K_L \frac{V^2}{2g} \quad . . . . \quad (4.14)$$

and K_L = the loss coefficient for the valve for the particular setting at a given time. Because K_L depends on time as determined by the closure scheme and the type of valve used, Equation 4.13 cannot be integrated directly and a numerical solution must be generated.

It should be noted that for sudden valve closure, dV/dt is infinite, and Equations 4.13 and 4.14 show that the pressure p_2 upstream of the valve also become infinite. Of course the actual pressure does not become infinite and the rigid water column theory gives meaningless results.

Complex Pipe Systems

Single constant-diameter pipes do not occur in most cases. However, techniques exist which permit complex systems of series and parallel pipes to be reduced to single, constant-diameter equivalent pipes. The equivalent pipe can then be analyzed by the techniques set forth earlier. Because rigid water column theory is used infrequently in light of the ease and availability of the application of elastic theory, complex pipe systems will not be addressed here. Rather, other references such as Watters [9] should be consulted for detailed information.

Application of Rigid Water Column Theory

Rigid water column theory should only be applied when changes of velocity are so slow as to cause a minimum of elastic effects. It was shown earlier that a suddenly closed valve can produce a pressure head increase of ΔH. In fact, if the valve were closed more slowly (but still completely) in less than 2 L/a seconds, the full pressure head increase ΔH would still be reached. Rigid water column theory would not apply in either case. To help grasp the differences between the two theories, Figure 8 was prepared showing the pressure increases caused by a linear decrease in velocity with time. In this case it appears that both theories give reasonably similar results if the required valve closure time is 10 L/a seconds or longer. Also note that rigid water column theory does not generally give conservative results.

4.2.3 Elastic Theory

It is now appropriate to examine the effects of fluid and conduit elasticity on the pressures generated by changes in velocity. Having observed what happens when the valve in a pipeline is suddenly closed, it remains now to develop some equations for calculating pressure from velocity changes. These equations must incorporate elastic effects so the equations we have used previously for rigid water column theory cannot be directly applied. However, the equations of impulse-momentum and conservation of mass along with the elastic deformation equations of the fluid and the pipe materials will be used.

The ΔH-Equation

First, apply the impulse-momentum equation to derive an expression for the change in head resulting from a local change in velocity. Realizing that a small change in velocity will cause a pressure wave to propagate upstream, consider a piece of cylindrical pipe ΔL long.

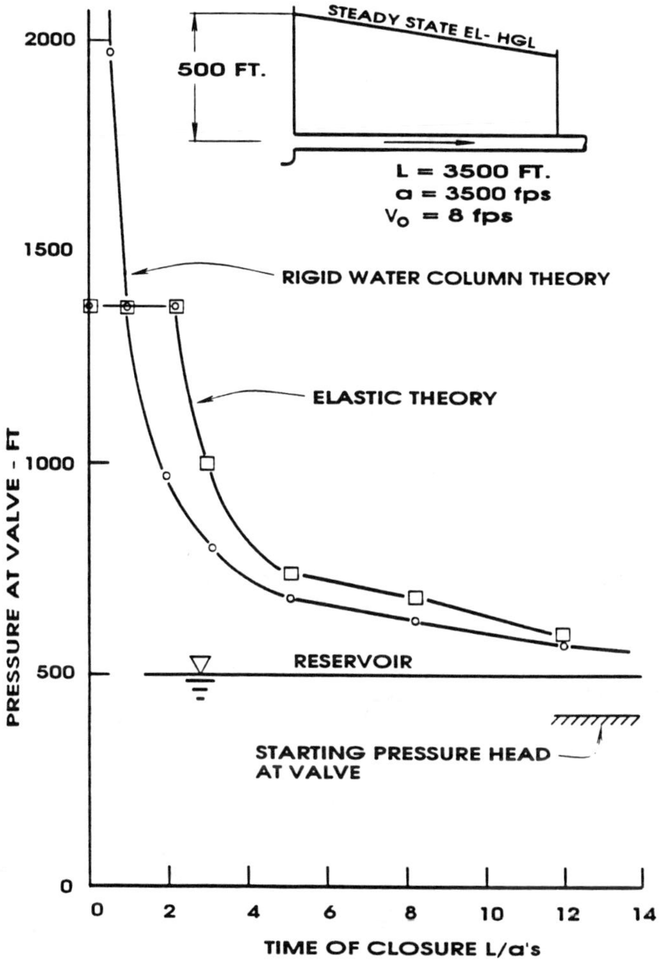

Figure 8.
Comparison between elastic theory and rigid water column theory for a range of valve closure times where velocity decreases linearly with time.

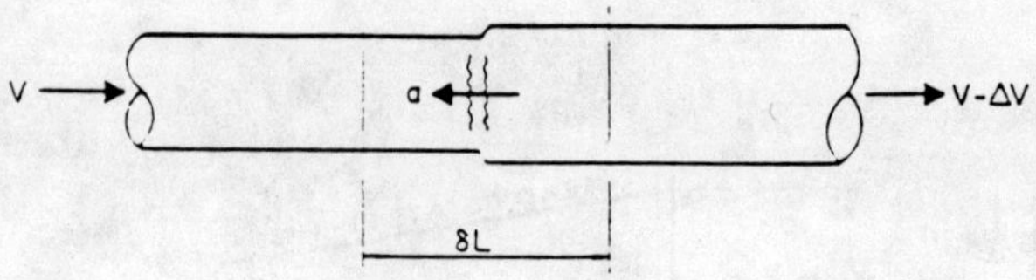

Unsteady Case

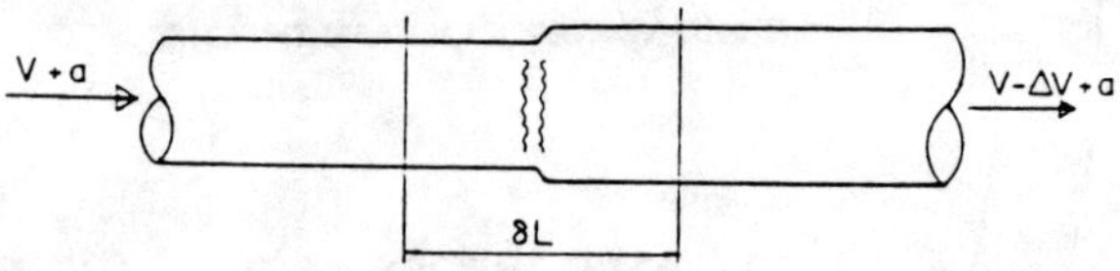

Steady Case

Figure 9
Converting the unsteady flow to a steady flow situation. (Watters [9]).

The pressure wave and bulge propagate at speed **a** however, moving with the pressure wave creates a "steady flow" situation which can be analyzed by elementary equations.

Applying the impulse-momentum equation to the "steady flow,"

$$(\Sigma F_{ext})_x = Q\rho\,(V_{out} - V_{in}) \quad \text{. . . . . . .} \quad (4.15)$$

where F_x = forces in the x-direction
Q = "steady state" discharge

Ignoring terms which have small effect, the following equations results.

$$\Delta H = -\frac{a}{g}\Delta V \quad \text{. . . . . . .} \quad (4.16)$$

Equation 4.16 is valid for relatively rigid pipes ranging in elasticity from steel and concrete down to PVC.

The **a**-Equation

It is still not possible to compute ΔH unless the wave speed **a** is known. To compute the wave speed it is necessary to apply principles of conservation of mass and to calculate elastic deformations of both pipe and fluid. The result of the analysis gives

$$a = \frac{[K/\rho]^{1/2}}{\left[1 + \frac{K}{E}\frac{D}{e}(c')\right]^{1/2}} \quad \text{. . . . . . .} \quad (4.17)$$

where K = bulk modulus of elasticity of the liquid, psi
E = modulus of elasticity of pipe material, psi
D = inside pipe diameter, in
e = pipe wall thickness, in
c' = coefficient which depends on pipe restraint

For water the numerator of Equation 4.17 is approximately 4720 ft/sec. Equation 4.17 should be applied only to fairly rigid pipes where E, K can be determined accurately and the relative thickness of the pipe wall D/e is greater than 40. For other situations, see Wylie and Streeter [12], Watters [9], or Chaudhry [5].

The values of c' can be calculated after establishing the type of pipe restraint. If the pipe is anchored throughout against axial movement, $c' = 1 - \mu^2$ where μ = Poisson's ratio. If the pipe has functioning expansion joints throughout its length, c' = 1; and if the pipe is supported only at one end and is allowed to stress and strain both laterally and longitudinally, $c' = 5/4 - \mu$. For reference, some values of Poisson's ratio and E are listed below:

units?

$$a = \frac{(3\times10^5/62.4)^{1/2}}{\left(1 + \frac{3\times10^5}{24\times10^6}\cdot\frac{0.5}{0.25}(1-.28^2)\right)^{1/2}} =$$

Steel	$E = 30 \times 10^6$ psi	$\mu \approx 0.30$
Ductile Iron	$E = 24 \times 10^6$ psi	$\mu \approx 0.28$
Copper	$E = 16 \times 10^6$ psi	$\mu \approx 0.36$
Brass	$E = 15 \times 10^6$ psi	$\mu \approx 0.34$
Aluminum	$E = 10.5 \times 10^6$ psi	$\mu \approx 0.33$
PVC	$E \approx 4 \times 10^5$ psi	$\mu \approx 0.45$
Fiberglass reinforced plastic (FRP)	$E_2 = 4.0 \times 10^6$ psi	$\mu_2 = 0.27 - 0.30$
	$E_1 = 1.3 \times 10^6$ psi	$\mu_1 = 0.20 - 0.24$
Asbestos Cement	$E \approx 3.4 \times 10^6$ psi	$\mu \approx 0.30$
Concrete	$E = 57{,}000 \sqrt{f'_c}$	$\mu \approx 0.24$ (dynamically)
where f'_c = 28-day strength.		

The value of K for water is generally taken as approximately 300,000 psi.

To get a "feel" for the magnitude of pressures which can develop under water hammer situations, consider an example.

Example 4.6

Suppose a 10,000 foot long steel pipe contains water flowing at 10 fps. If the wave speed in the pipe were 3220 fps, what increase in head would result if a valve at the downstream end were suddenly closed?

$$\Delta H \approx -\frac{a}{g}\Delta V = -\frac{3220}{32.2} \times (-10)$$

$$\Delta H \approx 1000 \text{ ft. or } 433 \text{ psi}$$

Also note that **L/a** = 10,000/3,220 = 3.11 seconds. Therefore, if the valve closed in less than 6.22 seconds, the full water hammer pressure of 433 psi would still be developed. Charts are available for estimating pressures generated when closing valves in times greater than 2 **L/a** [11]. However, computer analysis of each particular system is more accurate and generally no more difficult.

Computer Analysis

In the general analysis of a pipe system undergoing water hammer, a set of partial differential equations are solved to determine pressure head and velocity at any location in the system for any types of boundary conditions imposed. The equations come about from the application of Newton's second law and the conservation of mass to a section of fluid and pipe only **ds** long. The resulting equations are:

$$\frac{dV}{dt} + \frac{1}{\rho}\frac{\partial p}{\partial s} + g\frac{dz}{ds} + \frac{f}{2D}V|V| = 0 \quad . \ . \ . \ . \quad (4.18)$$

and

$$a^2\frac{\partial V}{\partial s} + \frac{1}{\rho}\frac{dp}{dt} = 0 \quad . \ . \ . \ . \ . \ . \ . \quad (4.19)$$

where s = location along the pipe
z = elevation above the datum

To solve these equations by modern computer methods, a procedure known as the method of characteristics is used. The two partial differential equations are converted to ordinary differential equations. The ordinary differential equations are represented in finite difference form. Then the resulting finite difference equations are solved by computer.

Once the basic computer program has been written, it is very easy to show the effects of friction, pipe profile, valve closure, etc. Furthermore, the basic computer program is not large nor does it require large storage capabilities in the computer. As a result, the programs can be executed on small digital computers or personal computers generally available in most engineering firms. A basic program is presented on the following page.

Effect of Friction

The effect of friction on water hammer pressures was rather complicated to access accurately before modern computer methods of analysis were developed. Although frictional effects are easily handled by computer, the following example illustrates an approximate method for estimating the effect of friction on high pressures in a simple case. Consider the system shown in Figure 10.

BASIC WATER HAMMER PROGRAM FOR A SIMPLE PIPE

```
      PROGRAM PROG4
C*********************************************************************
C                   PROGRAM NO. 4
C  APPROXIMATE WATER HAMMER PROGRAM FOR A SINGLE PIPE
C
C  THIS PROGRAM HAS BEEN INCLUDED FOR THE CONVENIENCE OF THE READER AND
C  THE AUTHOR ACCEPTS NO RESPONSIBILITY FOR ITS CORRECTNESS. USERS OF
C  THIS PROGRAM DO SO AT THEIR OWN RISK.
C*********************************************************************
C
C           -- RESERVOIR AT UPSTREAM END --
C  -- VELOCITY DECREASES LINEARLY WITH TIME TO ZERO AT DOWNSTREAM VALVE
C
C           *****  DATA DESCRIPTION  *****
C   TITLE = JOB DESCRIPTION. ANY INFORMATION UP TO 80 COLUMNS MAXIMUM
C   IOUT = PRINT OUTPUT INDEX. GIVES PRINTED OUTPUT EVERY IOUT-TH TIME
C          STEP. FOR EXAMPLE, IF IOUT=3 THEN OUTPUT IS PRINTED EVERY
C          THIRD TIME STEP.
C   NPARTS = NUMBER OF PIPE SEGMENTS INTO WHICH PIPE IS DIVIDED
C   D = PIPE DIAM-IN  L = PIPE LENGTH-FT
C   F = DARCY-WEISBACH F-VALUE OR HAZEN-WILLIAMS C-VALUE
C   A = WAVE SPEED-FPS  VZERO = INITIAL STEADY STATE VELOCITY-FPS
C   HZERO = UPSTEAM RESERVOIR ELEVATION-FT
C   ELEVUP = ELEVATION OF UPSTREAM END OF PIPE-FT
C   ELEVDN = ELEVATION OF DOWNSTREAM END OF PIPE-FT
C   TMAX = MAXIMUM REAL TIME OF SIMULATION-SEC
C   TCLOSE = TIME REQUIRED FOR VALVE CLOSURE-SEC
C
      DIMENSION X(20),V(20),H(20),HLOW(20),HHIGH(20),HEAD(20),
     $VNEW(20),HNEW(20),PIPEZ(20)
      REAL L,NEXP
      CHARACTER TITLE*80,ROUGH*3,CH,CH12,CH78,CH79,CHN
C
C---------------------------------------------------------------------
      WRITE(*,799)
      OPEN(5,FILE='      ')
      READ(5,100) TITLE
      READ(5,*) NPARTS,IOUT
      READ(5,*) D,L,F,A,VZERO
      READ(5,*) HZERO,ELEVUP,ELEVDN,TCLOSE,TMAX
      OPEN(6,FILE='',STATUS='NEW')
C---------------------------------------------------------------------
C
      CH=CHAR(27)
      CH12=CHAR(12)
      CH78=CHAR(78)
      CH79=CHAR(79)
      CHN=CHAR(3)
      WTT=L/A
      DELL=L/NPARTS
      T=0.
      NEXP=1.0
```

```
      ROUGH='F ='
      IF(F.GT.10.) NEXP=0.85
      IF(F.GT.10.) ROUGH='C ='
      DELT=DELL/A
      C=32.2/A
      INDEX=TMAX/DELT + 1
      DELEL=(ELEVDN-ELEVUP)/NPARTS
      NODES=NPARTS+1
      WRITE(6,101) CH,CH78,CHN
      WRITE(6,200)
      WRITE(6,203) TITLE
      WRITE(6,201) IOUT,NPARTS,L,A,D,ROUGH,F,VZERO,HZERO,ELEVUP,
     $ELEVDN,WTT,TCLOSE,TMAX,DELT
      AK=12.*F*DELT/(2.0*D)
      IF(F.GT.10.) AK=12.*DELT*195./(2.0*D*(F**1.85)*(D/12.)**.17)
      IF(F.GT.10.) F=195./((F**1.85)*(VZERO**.15)*((D/12.)**.17))
      DELHF=12.*F*DELL*VZERO**2/(64.4*D)
      DO 300 I=1,NODES
      V(I)=VZERO
      H(I)=HZERO-(I-1)*DELHF
      HLOW(I)=H(I)
      HHIGH(I)=H(I)
      X(I)=(I-1)*DELL/L
      PIPEZ(I)=ELEVUP+(I-1)*DELEL
      HEAD(I)=H(I)-PIPEZ(I)
  300 CONTINUE
      WRITE(6,101) CH12
      WRITE(6,202)
      WRITE(6,204) T,(X(I),HEAD(I),H(I),V(I),I=1,NODES)
      DO 99 II=1,INDEX
      T=T+DELT
C   ** COMPUTE H AND V AT INTERIOR NODES **
      DO 20 I=2,NPARTS
      VNEW(I)=0.5*(V(I-1)+V(I+1)+C*(H(I-1)-H(I+1))-AK*(V(I-1)*ABS(V(I-1)
     $)**NEXP+V(I+1)*ABS(V(I+1))**NEXP))
   20 HNEW(I)=0.5*(H(I-1)+H(I+1)+(V(I-1)-V(I+1))/C-AK*(V(I-1)*ABS(V(I-1)
     $)**NEXP-V(I+1)*ABS(V(I+1))**NEXP)/C)
C ** COMPUTE H AND V AT UPSTREAM END **
C     -- THIS BOUNDARY CONDITION IS FOR A CONSTANT-HEAD RESERVOIR --
      HNEW(1)=HZERO
      VNEW(1)=V(2)+C*(HNEW(1)-H(2))-AK*V(2)*ABS(V(2))**NEXP
C ** COMPUTE H AND V AT DOWNSTREAM END **
C     -- THIS BOUNDARY CONDITION IS FOR LINEARLY DECREASING VELOCITY --
      IF(T.GT.TCLOSE) GO TO 30
      VNEW(NODES)=VZERO*(1.-T/TCLOSE)
      GO TO 31
   30 VNEW(NODES)=0.0
   31 HNEW(NODES)=H(NPARTS)+(V(NPARTS)-VNEW(NODES)-AK*V(NPARTS)*
     $ABS(V(NPARTS))**NEXP)/C
      DO 50 I=1,NODES
      IF(HNEW(I).LT.HLOW(I)) HLOW(I)=HNEW(I)
      IF(HNEW(I).GT.HHIGH(I)) HHIGH(I)=HNEW(I)
```

```
   50 HEAD(I)=HNEW(I)-PIPEZ(I)
      IF(MOD(II,IOUT).EQ.0) WRITE(6,204) T,(X(I),HEAD(I),HNEW(I),VNEW(I)
     $,I=1,NODES)
      IF(T.GT.TMAX) GO TO 400
      DO 40 I=1,NODES
      V(I)=VNEW(I)
   40 H(I)=HNEW(I)
   99 CONTINUE
  400 WRITE(6,101) CH12
      WRITE(6,205)
      DO 401 I=1,NODES
      HEADMX=HHIGH(I)-PIPEZ(I)
      HEADMN=HLOW(I)-PIPEZ(I)
  401 WRITE(6,206) X(I),HEADMX,HEADMN,HHIGH(I),HLOW(I)
      WRITE(6,101) CH,CH79
C
  100 FORMAT(A)
  101 FORMAT(3A)
  200 FORMAT(///20X,33('*')/20X,'* WATER HAMMER IN A SIMPLE PIPE *'/
     $20X,33('*'))
  201 FORMAT(///29X,'INPUT DATA'/29X,10('-')//28X,'IOUT =',I3/26X,'NPART
     $S =',I3//31X,'L =',F7.1,' FT'/31X,'A =',F7.1,' FPS'/
     $31X,'D =',F7.2,' IN'/31X,A,F9.4//
     $27X,'VZERO =',F7.2,' FPS'/27X,'HZERO =',F7.1,' FT'/
     $26X,'ELEVUP =',F7.1,' FT'/26X,'ELEVDN =',F7.1,' FT'//
     $29X,'L/A =',F7.3,' SEC'//26X,'TCLOSE =',F7.2,' SEC'/
     $28X,'TMAX =',F7.2,' SEC'/28X,'DELT =',F7.3,' SEC')
  202 FORMAT(///'  PRESSURE HEADS, H-VALUES AND VELOCITIES AS FUNCTIONS
     $OF TIME'/2X,60('-'))
  203 FORMAT(//10X,A)
  204 FORMAT(//18X,2(4X,'  X   HEAD-FT  H-FT   V-FPS')/' TIME =',F7.3,
     $' SEC',2(4X,'----- ------- -----  ------')/(18X,2(4X,F5.3,2F7.0,
     $F8.2)))
  205 FORMAT(////18X,27('*')/18X,'* TABLE OF EXTREME VALUES *'/18X,27('*
     $')//13X,'X    MAX HEAD  MIN HEAD   MAX H   MIN H'/11X,5('-'),2X,8(
     $'-'),2X,8('-'),2X,6('-'),2X,6('-'))
  206 FORMAT(11X,F5.3,2X,F7.0,3X,F7.0,3X,F6.0,2X,F6.0)
  799 FORMAT(//' YOU WILL BE REQUESTED TO SUPPLY THE NAMES OF YOUR'/
     $' INPUT DATA FILE (UNIT 5) AND YOUR OUTPUT FILE (UNIT 6).'/
     $' RESPOND TO THE PROMPT (?) WITH THE DISK DRIVE AND FILE NAME'/
     $' FOR EXAMPLE "C:DATA6"'//)
      END
```

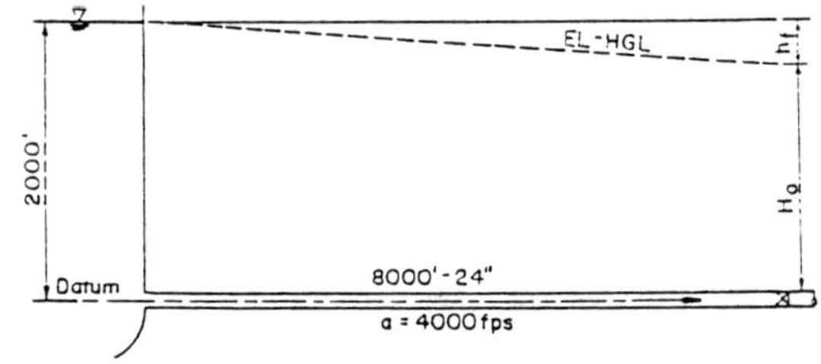

Figure 10.
Steady state configuration for water hammer with friction (Watters [9]).

The frictional head loss under steady state conditions is h_f. For sudden valve closure, analyses have shown that the maximum pressure head at the valve is approximately $H_{valve} + \Delta H + h_f$.

To understand this result and to develop a feel for frictional effects in general, refer to Figure 11. This figure represents the successive location of the EL-HGL after sudden valve closure in a simple pipe system. As the progress of the pressure waves is followed through the system, note how the effects of friction opposing the flow alter the EL-HGL position from that shown in Figure 5. Also note in zero-velocity portions of the pipe that small adjustments in EL-HGL slope are continually being made in an attempt to make the EL-HGL horizontal.

Pressure Reflection and Transmission at Junctions

When a water hammer wave moves up a small pipe and enters a larger pipe, there is the question of what pressure is transmitted into the larger pipe and what happens to the pressure situation in the smaller pipe.

An analysis of the situation using elastic theory and neglecting local friction reveals that the amount ΔH_1 passed on to the larger pipe is

$$\Delta H_1 = \frac{2A_2}{A_1 + A_2}\Delta H$$

for the series pipe and

$$\Delta H_1 = \Delta H_2 = \frac{2A_3}{2A + A_3}\Delta H$$

for the tee connection.

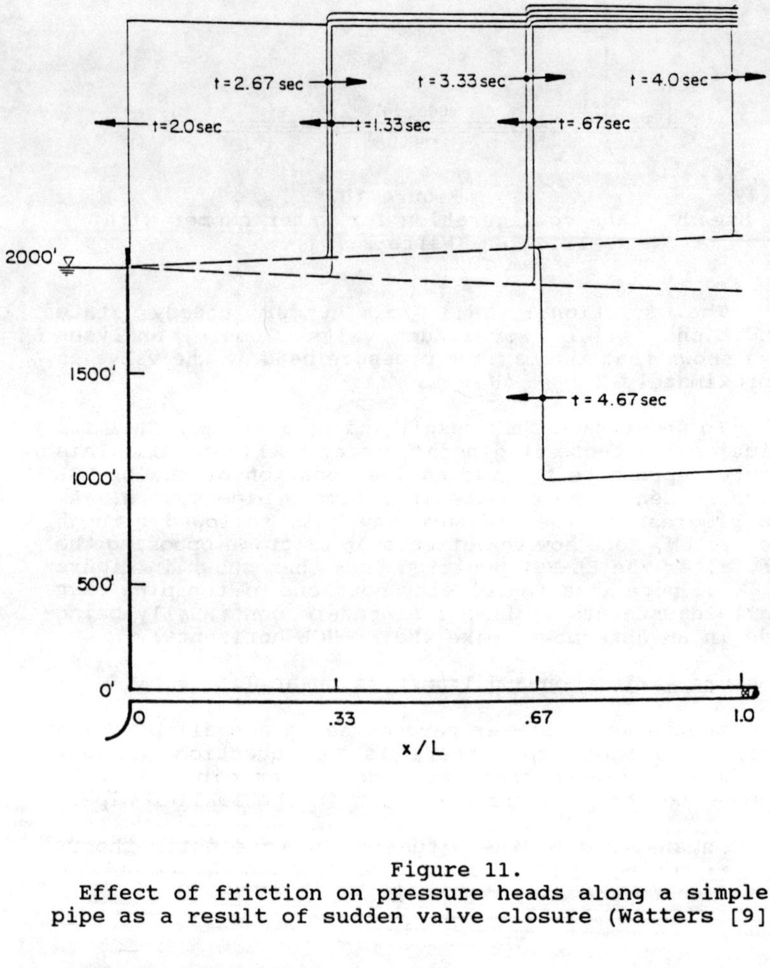

Figure 11.
Effect of friction on pressure heads along a simple pipe as a result of sudden valve closure (Watters [9]).

The wave speed is assumed to be the same in all pipes. The pressure head in the small pipe will reduce to match that in the large pipe upon reflection.

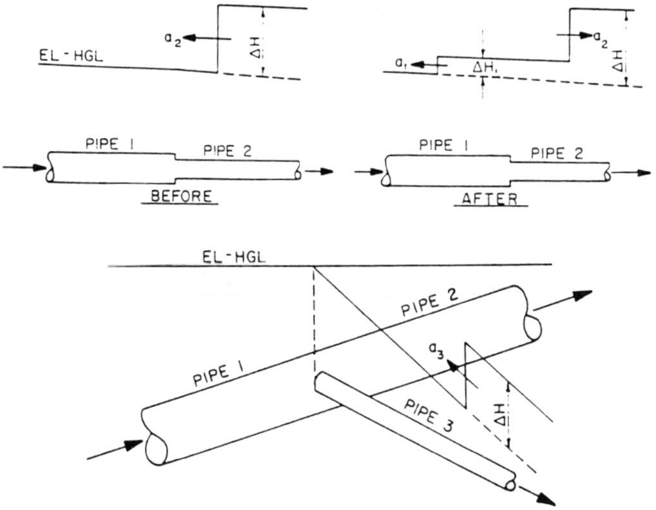

For example, if D = 6 inches, D_1 = 24 inches and ΔH = 500 feet, the portion of ΔH transmitted into the series pipe is 59 feet and into the tee connection is 30 feet. It is clear that the larger pipe acts much as a reservoir.

Dead-End Pipes

If a pipe system contains a member which carries no flow and terminates in a dead end, for example, a closed valve, then a unique situation exists which could cause unexpected high pressures. This is actually a special case of the tee junction situation discussed previously. As a high pressure wave passes the junction from which the dead-end pipe extends, a high pressure wave moves into the pipe and induces a flow velocity toward the dead end. When the pressure wave and the induced velocity reach the dead end, the flow is abruptly stopped, thereby increasing the head at the dead end two increments in pressure.

While pipe system geometry, pipe sizes, and hydraulic friction losses all affect the total pressure increase to some extent, the maximum pressure increase occurs when the dead-end pipe is very small in relation to the main pipe. Under this condition and with small frictional effects, the pressure increase in the dead-end pipe is, at most, equal to twice the value of the pressure head increase of the wave initially passing the junction.

Experimental Verification of Theory

The laboratory verification of the validity of the classical theory as applied to water hammer situations is well established. For example, work by Watters, Jeppson, and Flammer [10] illustrates how PVC pipe and reinforced plastic pipe behave under water hammer action. This work, supported by the Johns-Manville Corporation, shows how plastic pipes can be handled in water hammer situations both in above-ground and buried situations. The experiments showed that wave speed could be calculated to acceptable accuracy using classical formulas provided the buried pipe was assumed to be physically constrained. For these pipes a comparison of theoretical and actual pressure is presented in Figure 12 for PVC pipe. In general, these figures show that the classical analysis is verified.

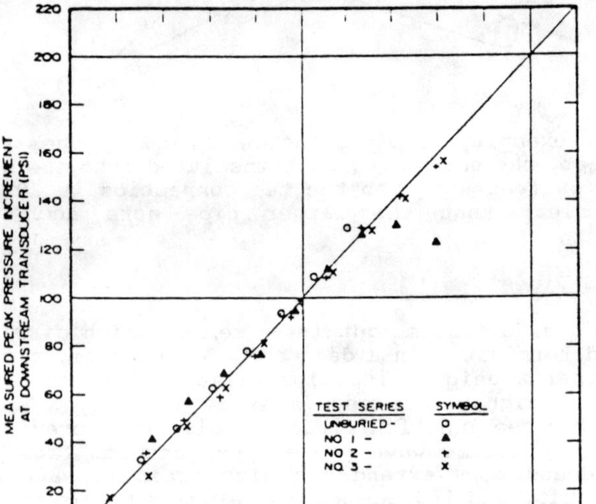

Figure 12.
Comparison of computer and experimental pressure increments in PVC pipe.
(Watters, Jeppson and Flammer [10]).

An example of experimental verification in a constructed and tested pipeline system is shown in Figure 13. This long pipeline with several surge tanks was designed using classical theory and field tested after construction. There are discrepancies but no problems which would have endangered the system.

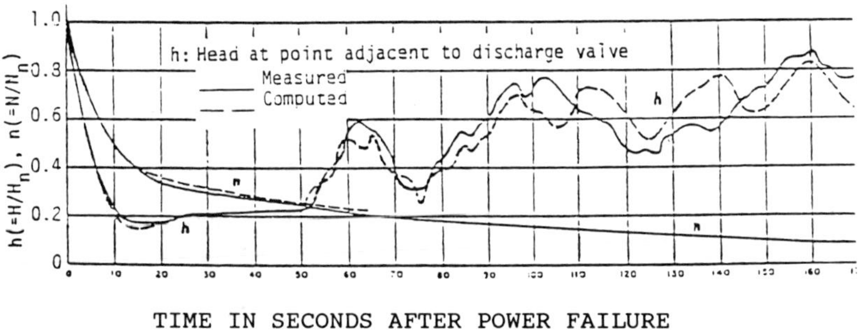

Figure 13.
Hydraulic transients measured in field tests after power failure to the pumps
(from Miyashiro, Trans. ASME Oct. 1967).

4.2.4 Some Causes of Water Hammer Problems

Water hammer problems result from both excessively high and excessively low pressures. Excessively high pressures can result in damage to pipeline appurtenances and pipe rupture. Low pressures can lead to column separation and, in some cases, collapse of the pipe. Column separation is the condition resulting when the pressure in the flow drops to the vapor pressure of the liquid and the liquid begins to vaporize, i.e., boil. Figure 14 illustrates a typical situation which can cause column separation. When the valve in the pipe is closed suddenly, the pressure upstream increases by ΔH and the pressure downstream tries to decrease by ΔH. However, in no instance may the pressure drop below the vapor pressure of the liquid (for water at sea level, $H_{atm} \approx 33$ ft); hence, the desired pressure drop cannot occur. In order to bring the water just downstream of the suddenly closed valve to rest, a pressure drop of ΔH must occur. Because this pressure head drop is physically impossible, the water next to the valve moves away leaving a cavity filled with water vapor. If this condition does not cause the pipe to collapse, damage can still be done when reverse flow occurs, causing the cavity to collapse with a shock as the water column collides with the valve. Under certain conditions, high shock pressures can result which can cause pipe rupture. In brittle pipes or cement-lined pipes, even if pressure is not initially high enough to rupture the pipe, the shock can cause fracture or spalling of the lining or the steel reinforcing protective cover and, ultimately, pipe failure.

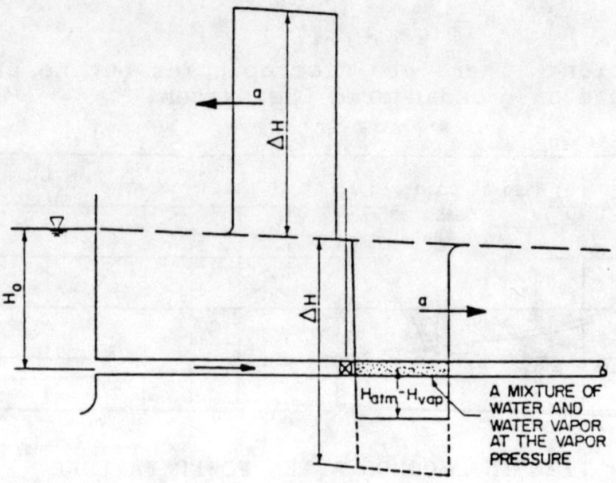

Figure 14.
Column separation caused by sudden valve closure (Watters [9]).

Valve Closure or Opening

Valve movement, whether opening or closing, causes velocity changes which result in pressure changes. As already seen in Figure 14, valve closure can lead to low-pressure problems as well as high-pressure problems. What is not so obvious is that column separation can occur next to the upstream side of a suddenly closed valve. As shown on Figure 6 for a horizontal pipe with a reservoir upstream, the positive ΔH wave drops to a negative ΔH wave at 2 **L/a** seconds. Should this drop in pressure reach the vapor pressure of the liquid, then column separation would begin upstream of the valve 2 **L/a** seconds after sudden valve closure.

There is another very significant effect of valve closure on pressure increase which should be recognized. It has already been shown that in a simple pipe, if the valve is closed in a time less than 2 **L/A** seconds, then the full water hammer pressure is developed. Intuition would suggest that if the valve were closed in 4 **L/A** seconds, a significant reduction in pressure would be realized. In general this does not happen. Examination of Figure 15 reveals some very interesting results. To interpret the results, it is necessary to recall the function of the valve.

The valve slows down the flow by creating head loss in the system. If there is a large amount of head available in the system then the valve must be closed to the point where substantial head loss is generated before a significant drop in velocity is experienced. In the case of a gate valve, this does not occur until the valve is 95 to 98 percent closed. Consequently, in most cases, the schedule for closing the valve the final 2 to 5 percent is the most critical in determining the pressure increase generated.

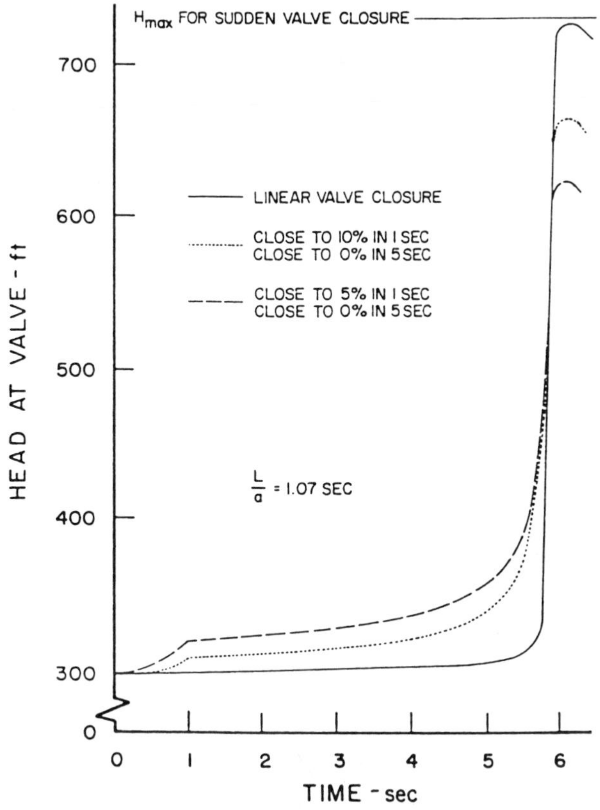

Figure 15.
Water hammer resulting from various gate valve closure schemes (Watters [9]).

<u>Check Valves</u>

Check valves can cause large water hammer pressures if a flow backwards through them must occur in order to close them. In this situation, the slamming check valve creates the same situation previously examined for sudden valve closure. Most modern check valves are of the "non-slam" variety. The mechanism for closure is generally spring or weight loaded in such a manner as to close the check valve at the instant forward flow ceases. Specification of non-slam check valves will minimize this source of water hammer problems.

Air Exhaustion

Filling empty lines, particularly in pumped systems, can give velocities well in excess of expected steady state velocities. At the low head generally occurring during the filling process, the pump is operating at a point on its characteristic curve where large flow rates occur. If the line is terminated by any device acting as a flow obstruction, e.g., a partially-closed valve or closed valve with an air-vacuum valve upstream, then a serious water hammer situation can develop (see Figure 16).

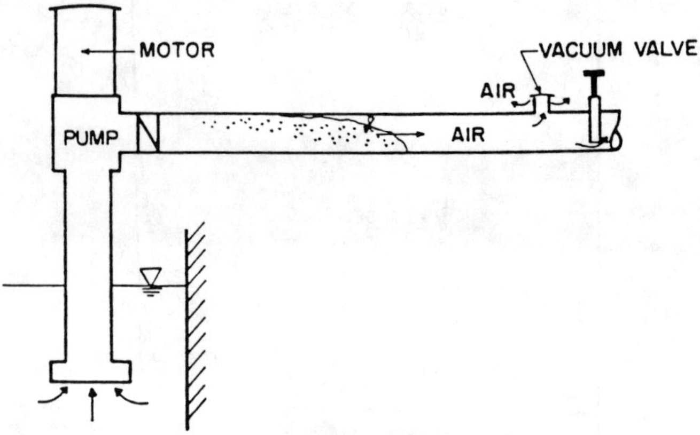

Figure 16.
Filling an initially empty pump discharge line (Watters [9]).

The air rushing out of the pipe ahead of the water will leave the pipe more easily than the water behind it. As a result, when the last air leaves and the water tries to escape, sudden decreases in velocity generally occur causing large water hammer pressures.

Another situation is depicted in Figure 17 where the discharge column of a pump is initially empty and the pipeline full of water is held by a closed check valve.

Large pressures will occur when the rushing water fills

the column and slams shut the vacuum valve.

An unusual set of circumstances can occur which can produce severe water hammer pressures. If power fails (or pumps are shut off) and the resulting drop in pressure creates air or vapor cavities in the pipeline, re-starting the pumps too soon could greatly aggravate the air exhaustion or cavity closure problem. Having the pumps come back on line creates higher cavity closure velocities which, in turn, could create pressures well beyond the allowable surge pressure in the pipeline.

Pump Starting in a Pipeline

As a pump in a pipeline with air removed is started up and comes "on line" a positive surge is generated in the pipe system. The magnitude of the surge is dependent on the sudden increase in discharge experienced when the pump forces open the check valve and begins to move the liquid. Without air in the line, the pressure increase created is generally not large. However, for pumps with high shutoff head, an objectionably high pressure might be created.

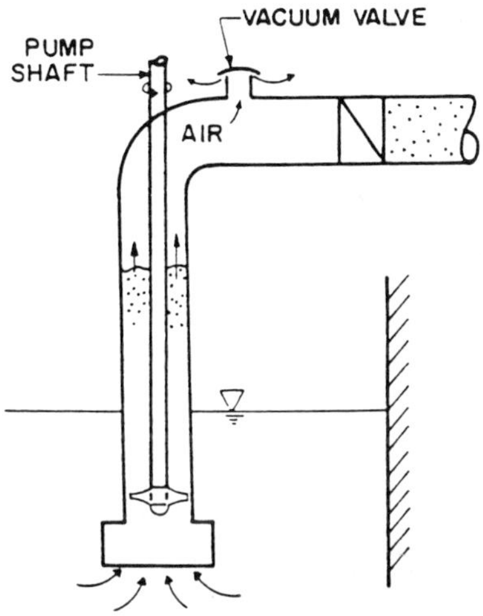

Figure 17.
Filling a pump discharge column behind a closed check valve (Watters [9]).

Power Failure (or Power Shutoff) to a Pump

If power is cut off from the pump motors suddenly, either accidentally or deliberately, significant water hammer problems can develop. Generally, the pressure drops rapidly downstream of the pump and this pressure drop propagates downstream at wave speed (see Figure 18). The drop in pressure can lead to extensive column separation and the consequences of cavity-closure shock could be severe.

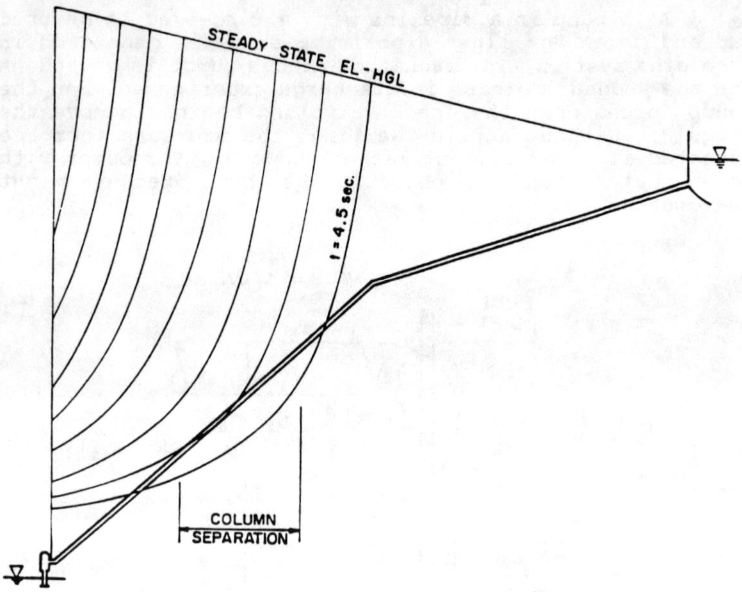

Figure 18.
Propagation of negative wave resulting from pump power failure (Watters [9]).

In addition, flow reversal in the system, if not properly handled, could lead to significant overpressures in the system generally in the vicinity of the pump.

4.2.5 Devices and Techniques for Controlling Pipeline Transients

There is one basic principle in transient analysis

and pipeline design to keep in mind. Transient conditions are related to changes in velocity. The change in pressure is directly proportional to the change in velocity. Avoid sudden velocity changes and you will generally avoid serious water hammer pressures.

Controlled Valve Movement

Earlier discussion illustrated how the valve closing schedule could affect the maximum pressures developed. It was shown that generally, the last 2 to 5 percent of valve closure is the most critical for gate valves. For other valves, other results will be discovered. The best way to determine the results of a valve closure schedule on water hammer pressures is to obtain the head loss characteristics of the valve and run a computer analysis. Within a relatively short time and with a small investment in computer time, the maximum pressures can be found and an acceptable closure scheme can be designed.

Once the proper valve closure schedule has been determined, a hand or motor-controlled system must be devised to operate the valve. Possibilities for control systems range from relatively simple staged motor-driven assemblies to rather exotic servo-mechanism devices.

Surge Relief Valves

On occasion it is necessary to close valves rapidly or create other obstructions in the flow which can cause sharp decreases in velocity and the resulting high water hammer pressures. In these cases, often the most economical solution is a surge relief valve. These valves, designed to open when a certain prescribed pressure is exceeded, range from relatively inexpensive spring-loaded devices, to rather expensive and complicated systems.

The surge relief valve is generally located on the side of the device expected to experience the high pressure. The purpose of the valve is to provide an escape for the onrushing liquid and prevent sudden decreases in velocity and the resulting high pressures. A high-quality surge relief valve has little inertia so that it opens almost instantaneously. It also must close in a manner which minimizes the loss of pipe liquid and yet avoids unnecessarily high pressures during the closure process. These requirements can lead to a rather expensive valve which must be field adjusted for proper performance.

To demonstrate the effectiveness of surge relief valves, consider the following example.

Example 4.7

A 36-inch pipeline 4200 feet long carries water at a velocity of 3.5 fps between two reservoirs (see the following sketch). The valve at the downstream end of the pipeline is closed and an 8-inch surge valve is set to open fully when the line pressure exceeds 130 psi. The pressure waves created both with and without the surge valve are shown on the sketch. It is clear that a substantial reduction in pressures is realized with the surge valve in operation.

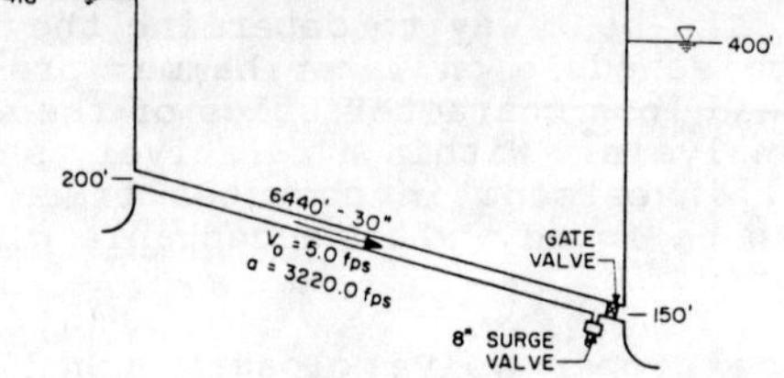

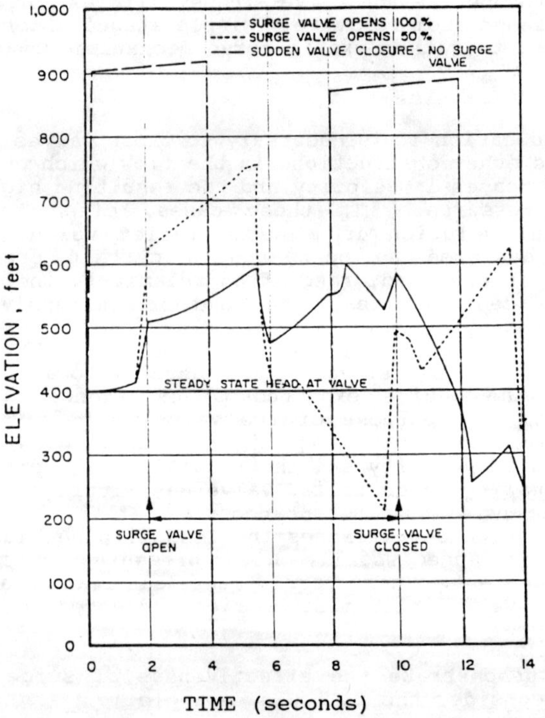

Non-Slam Check Valves

The problem of high pressure created by "slamming" check valves was discussed previously. The solution of specifying non-slam type, spring-loaded check valves was also presented so no further discussion is necessary.

Caution in Filling Empty Lines

The key to filling the empty lines of a pipeline system is to proceed with "caution." Means must be provided so that the liquid can be introduced slowly into the system at velocities of 1.0 fps or less. Air-release and air-vacuum valves must be provided so that the air can be forced from the system slowly. Valves must be provided at the outlets of the line (generally) so the line can be pressurized and air pockets forced out.

In pumped pipelines it may be necessary to provide a discharge by-pass to the sump to prevent objectionably high heads from occurring at the pump for low discharges. This requirement is almost always needed for axial-flow type pumps.

If it is necessary to fill a line with the pump "wide open," be sure there are no obstructions to cause sudden changes in velocity. Remember, if there are no sudden changes in velocity, there are generally no sudden changes in pressure.

Surge Tanks

Surge tanks can be employed to prevent both objectionably high and low pressures. In general they act as temporary storage devices for excess liquid which has been diverted from the main flow system to prevent high pressures; or they act as supplies of liquid to a pipe which needs more fluid to prevent excessive deceleration. The open-end surge tank can be used for both functions. In fact a single open-end surge tank can be so designed as to act effectively in controlling both high and low pressures in a pipeline. Figure 19 illustrates a typical application of an open-end surge tank in a hydroelectric installation where its purpose is to both store or supply water quickly as electrical load varies and to dampen the resulting pressure oscillations rapidly.

In pumped pipelines, the one-way surge tank is commonly used. Its purpose is to prevent column separation because it has no capability to prevent overpressures. Because its liquid surface is generally located well below the steady state hydraulic grade line, the one-way surge tank is isolated from the system with a check valve (see Figure 20). When the hydraulic grade line drops below the surface of the liquid in the one-way surge tank, the check valve opens and liquid flows into the system. The result is that the pipeline flow is not required to decelerate so rapidly and the hydraulic grade line at the location of the one-way surge tank is fixed close to the water surface in the tank. The resulting hydraulic grade line configuration for a typical pumped pipeline situation is shown in Figure 21.

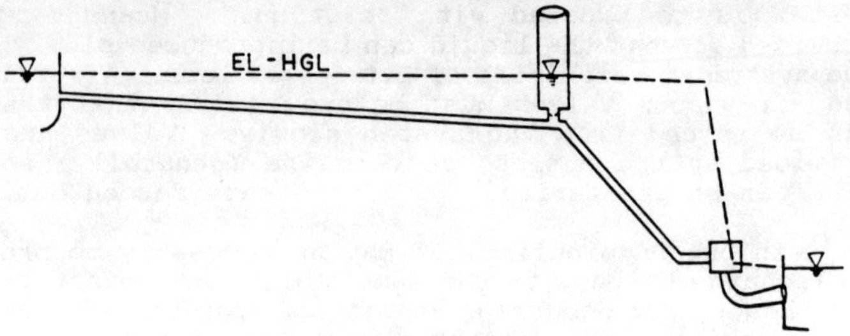

Figure 19.
Open-end surge tank in a hydroelectric installation (Watters [9]).

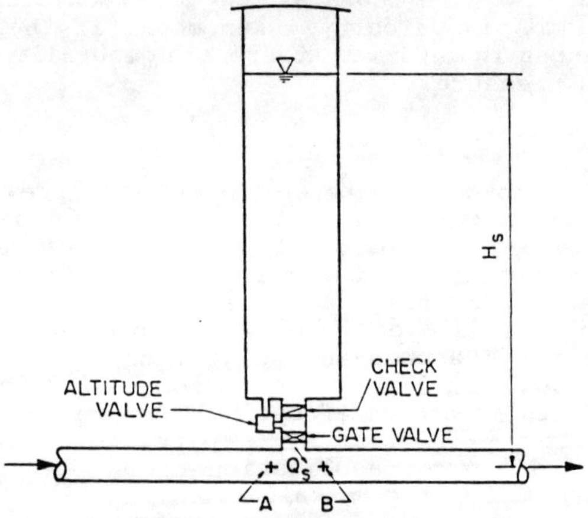

Figure 20.
A schematic sketch of a one-way surge tank (Watters [9]).

Air Chamber

The primary purpose of the air chamber is to prevent negative pressures and column separation in the pipeline system downstream of the chamber. However, the device is an excellent surge suppressor as well. The chamber is sealed and a layer of compressed air overlays the liquid inside (see Figure 22). After pump power failure, liquid is drawn from the chamber into the pipeline and the volume of air within the chamber expands, causing the pressure to drop. The rate at which the chamber pressure drops is dependent on the initial air volume, the thermodynamic process which the air undergoes, and the rate at which liquid is being drawn out of the chamber. With the chamber supplying liquid to the pipe at ever-decreasing pressures, the liquid in the pipeline is slowly decelerated without causing unnecessarily low pressures. By this means column separation can often be entirely avoided. A typical situation is shown in Figure 23 which illustrates how gently the air chamber can bring the system to rest. Compare this result with Figure 18.

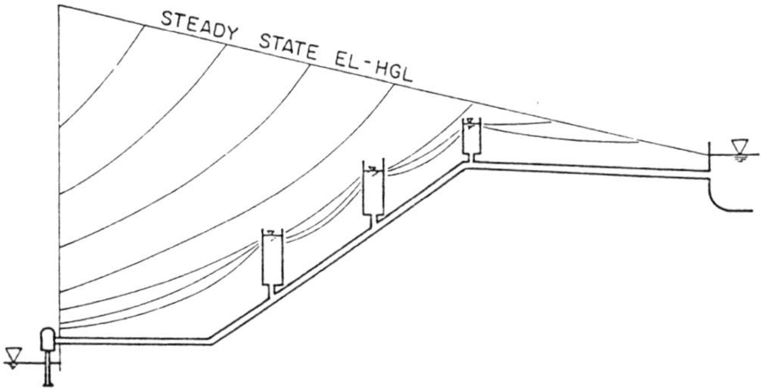

Figure 21.
One-way surge tanks in a pumped pipeline (Watters [9]).

Figure 23 also shows that an air chamber is not always adequate to completely prevent column separation. Low pressure can occur at local summits in the pipeline. The air chamber often cannot help enough at these locations. In these cases a one-way surge tank is provided at the summit to "drape" the EL-HGL above the pipe as shown in Figure 24.

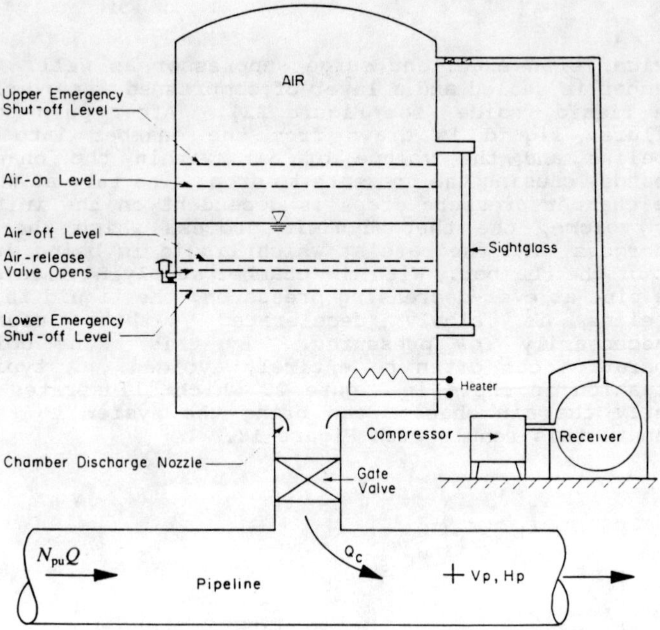

Figure 22.
Schematic diagram of an air chamber and its appurtenances (Watters [9]).

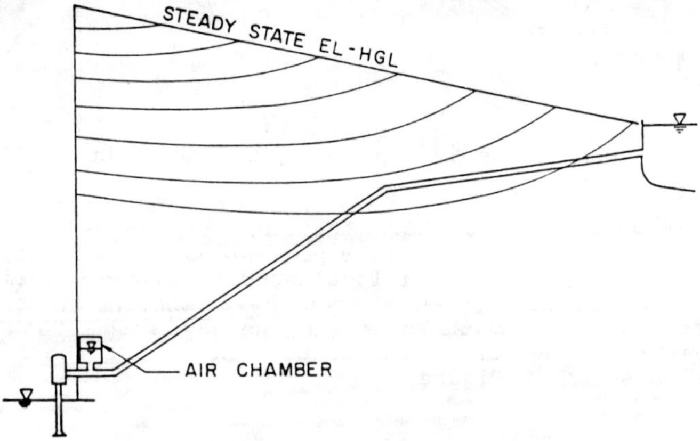

Figure 23.
Propagation of negative wave after pump power failure with an air chamber at the pump (Watters [9]).

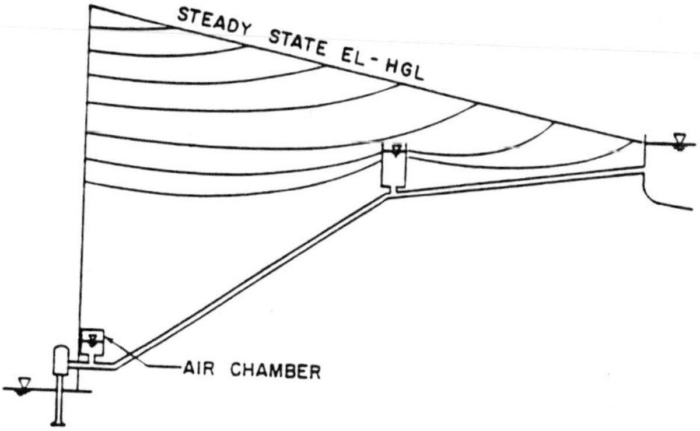

Figure 24.
Propagation of negative wave after pump power failure with air chamber and one-way surge tank (Watters [9]).

Example 4.8

A 6000 foot 8-inch pipeline leads from a pump station at 4700 feet of elevation to a reservoir at 5120 feet of elevation. Can column separation be expected under power failure conditions and, if so, what size air chamber is required to prevent it?

The following sketch shows that column separation does occur at approximately 1.5 seconds after power failure. A sequence of analyses with various sized air chambers reveals that one with an initial air volume of 25 ft^3 will prevent negative pressures from occurring. An air chamber 3 feet in diameter and 6 feet long was selected.

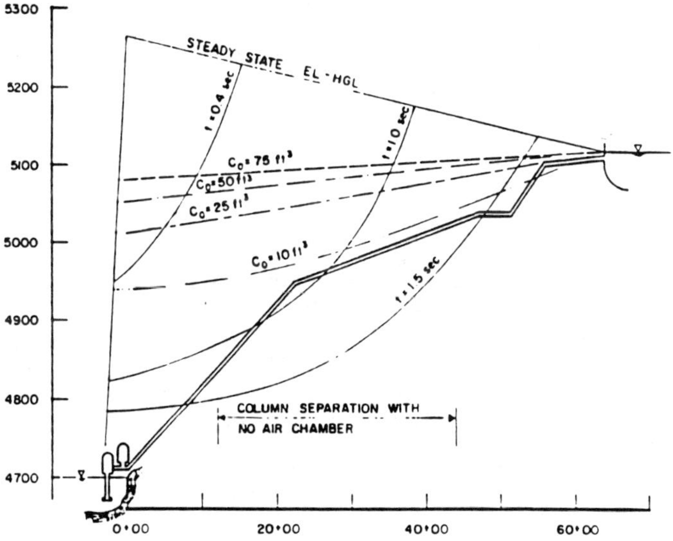

Vented Surge Tank

In special circumstances, a combination of the features of the one-way surge tank and the air chamber can be incorporated in a device called a vented surge tank (also referred to as a two-way surge tank). During downsurge, the device, an enclosed pressure tank, acts like a one-way surge tank, letting water into the line. As water is drained from the tank, air enters through a check valve opening to the atmosphere. There is no check valve in the line connecting the tank to the pipeline. When flow reverses and enters the tank, the air check valve closes, trapping the air and creating an air chamber environment. Air is slowly bled from the tank through an air release valve after the surge has been damped out. The device offers more versatility than a one-way surge tank and is more economical than an air chamber. Used properly, it can be an economical and effective means of surge control.

In-line Surge Suppression

Under the right conditions, the vented surge tank attributes can be realized within the pipeline. At a local summit or other suitable high location in the pipeline, vacuum valves and air release valves are located and sized as would be done for a vented surge tank. On downsurge, air enters the line holding the local pressure near atmospheric and creating an air pocket at the high point. On upsurge the vacuum check valves close, trapping the air and creating the air chamber effect. In essence, the pipeline in the vicinity of the summit has become a vented surge tank. Under the right conditions, this technique provides a simple and very economical means of surge control.

Surge Anticipation Valve

In certain cases, a device other than an air chamber can be used to prevent the high pressures caused by reverse flow or vapor cavity closure resulting from column separation after pump power failure. This device is appropriate if the pipe system can stand vacuum and if small amounts of air entering air-vacuum valves are not objectionable. The device is a surge relief valve which is set to open when pressure on the downstream side of the pump drops after power failure. Depending on the pipeline profile and other considerations, column separation may or may not occur after power failure. In any case, when the system, by virtue of gravity or pressure gradient, begins to flow back toward the pump, the surge valve is already open and no sudden deceleration occurs. This device is called a "surge anticipation valve." Keep in mind this valve does not prevent column separation. It only protects the system from high pressures caused by cavity closure or reverse flow against a closing or closed check or line valve. It has the additional desirable feature of being lower in cost and relatively maintenance-free when compared to the air chamber.

Summary

In summary, water hammer can lead to severe pressures in a pipeline system which can, in turn, cause the system to fail. There are a great many techniques and devices which can be employed to prevent serious problems from occurring. However, indiscriminate installation of the various control devices currently available from vendors can be a very costly approach to design which does not necessarily improve the safety of the pipeline system. A thorough analysis by a knowledgeable engineer, well-versed in transient analysis techniques, is a desirable design procedure in any pipeline system.

4.2.6 Guidelines for Water Hammer Investigations

Although transients occur in virtually all pipelines, it is not always necessary to analyze every situation. The following guidelines, taken from Sanks, et al [8], are presented to assist the designer in deciding what action is necessary.

Situations Not Requiring Analysis

1. Pumped pipelines with flows less than 100 gpm usually have low velocities and occur in small pipes with high pressure ratings. Transient pressures are small and the pipe is strong enough to contain the pressures.

2. Pipelines with velocities less than 2 fps generally cannot generate transient pressure high enough to exceed the allowable surge pressure of the pipe.

3. Transient pressures occurring in pipe networks are dissipated so quickly at branches and junctions that they are of only very local consequence.

4. Pumped pipelines with a pump head increase of less than 30 ft. generally do not have the potential of creating significant transient pressures.

Situations Requiring Analysis

1. Pumped pipelines with pump head increases greater than 50 ft. and with flows greater than 500 gpm have the potential of creating significant transient pressures.

2. Pumping installations with high static lift could experience high transient pressures if check valve slam occurs.

3. Any pipeline operated in such a manner as to cause air to be ingested into the system, can be subjected to severe pressures resulting from air exhaustion after pump restart.

4. Pipelines with profiles such that column separation may occur (See Fig. 18) may experience significant surge pressures resulting from cavity closure shock.

Conditions Which Suggest Serious Water Hammer Potential

The previous edition of this manual and Sanks, et al [8] provide a checklist of conditions which could cause surge problems. If one or more of these conditions occur, a serious problem may exist.

1. High summits in the pipeline.

2. Steep pipe gradients followed by relatively flat gradients.

3. Flow velocities in excess of 4 fps.

4. Factors of safety of the pipe less than 3.5 based on the normal operating pressure.

5. Flow reversal in less than t_c.

6. Check valve closure in less than t_c.

7. Any valve closure or opening in less than 10 sec.

8. Pump started with discharge valve open.

9. Booster stations that depend on operation of the main pumping station.

10. Presence of quick-closing automatic valves that become inoperative if power fails or pumping system pressure fails.

11. Pumps which restart automatically less than t_c after shutdown.

The decision as to whether or not to include a transient analysis as part of the design process is both an economic decision and one based on the above considerations. Because it is not generally possible to determine the full extent of water hammer problems without some level of analysis, it is recommended that a conservative approach be taken. That is, proceed far enough with the analysis to determine whether or not a significant water hammer condition may exist which would justify a thorough analysis.

REFERENCES

1. Cross, Hardy, Experiment Station Bulletin No. 286, University of Illinois, urbana, 1936.

2. Kerr. S. Logan, "Effect of Valve Operation on Water Hammer," Journal, Amreican Water Works Association, Vol. 52 (1960), No.1.

3. Parmakian, A., Waterhammer Analysis, Dover Publications, New York, 1963.

4. Streeter, Victor L., Waterhammer Analysis of Distribution Systems, 1P-748, University of Michigan, Ann Arbor, Oct. 1966.

5. Chaudhry, M.H., Applied Hydraulic Transients, Second Edition, Van Nostrand Reinhold, New York, 1987.

6. Hydraulic Institute Standards for Centrifugal, Rotary & Reciprocating Pumps, 14th Edition, Hydraulics Institute, Cleveland, 1983.

7. Karassik, I.J., W. Krutzsch, W. Fraser, and J. Messina, Pump Handbook, Second Edition, McGraw-Hill Book Co., New York, 1986.

8. Sanks, R.L., G. Tchobanoglous, D. Newton, B. Bosserman II, and G. Jones, Pumping Station Design, Butterworth Publishers, Boston, 1989.

9. Watters, G.Z., Analysis and Control of Unsteady Flow in Pipelines, Second Edition, Butterworth Publishers, Boston, 1984.

10. Watters, G., R. Jeppson and G. Flammer, "Water Hammer in PVC and Reinforced Plastic Pipe," Journal of Hydraulics Division, ASCE, V102, No.7, 1976.

11. Wood, D. and S. Jones, "Water Hammer Charts for Various Types of Valves," Journal of Hydraulics Division, ASCE, V99, No.1, 1973.

12. Wylie, E.B. and V.L. Streeter, Fluid Transients, McGraw-Hill Book Co., New York, 1978.

13. Hydraulic Institute, Hydraulic Institute Standards. 13th ed., Cleveland, Ohio.

SECTION 5 - MAINS

5.1 Purpose

The purpose of this section is to present recommendations on the design of water transmission mains and distribution pipelines and sewage force mains.

5.2 General

Transmission mains are large diameter pressure pipelines which convey water to the distribution system pipelines. Generally transmission mains are pipelines larger than 16-inches in diameter while distribution pipelines are 16-inches and smaller. Typically transmission and distribution pipelines are sized utilizing a flow network computer analysis of the overall system including supply and storage facilities and pressure reducing-pressure sustaining valves for multi-pressure zone systems. The flow network computer analysis should model conditions critical to system operation including peak hour and peak day plus fire flows. Extended period flow simulations which model varying flow demands, over 24 to 72 hours or more, are helpful in determining overall system operation including treatment facility capacity, tank and reservoir performance and volume requirements.

Acceptable headloss criteria for line sizing often varies between transmission and distribution pipelines with lower headlosses required for transmission pipelines. In addition, many utilities place limitations upon flow velocities in the pipeline during peak flow conditions.

Individual customer service taps to transmission pipelines are usually prohibited to permit multi-pressure zone systems and to facilitate operation and maintenance of the line.

Depth of bury of pipelines is a function of the probability of line protection from construction damage, surface live loads, freezing and rupture, scour depth at river crossings and occasionally, pipeline grades.

Sewage force mains are pressure pipelines from wastewater pump stations to a gravity line or treatment facility. Operating pressures and velocities are typically lower that water pressure pipelines.

5.3 Design Loads

Design conditions for pipelines include both internal pressures and external loads on the pipeline structure. The pipeline structure includes the support provided by the bedding surrounding the pipe.

5.3.1 Internal Pressures

Internal pressures on pipelines include both static and dynamic pressures. Static pressures are relatively constant over time. Dynamic pressures, or surges, are transient and occur when the flow velocity or pressure changes; such as when a valve is closed in the system, or a pump started or stopped. Surges may also result from entrapped air or liquid column separation.

Generally, water distribution system pipeline pressures vary from a minimum of 40 psi to a maximum of approximately 100 psi measured at the service tap. Lower pressures generally result in unacceptable levels of service in the residence or building when demands due to multi-uses of water occur. Often minimum system pressure is a function of required fire flows and residual pressures at hydrants in the distribution system. Higher pressures can be unacceptable due to design and operation limitations on domestic hot water heaters and pressure surges which frequently occur in domestic plumbing pipelines. For example, surges occur when valves close during operation cycles on washing machines and dishwashers which may damage appliances or cause the water heater temperature and pressure relief valve to open. Higher levels of plumbing noise are present with higher pressures.

Transmission line pressures are frequently greater than 100 psi. Transmission mains frequently pass through and serve several pressure zones. Typically, these zones are fed from the higher pressure transmission main through pressure reducing valves. Also, a single source of supply in a multi-pressure zone water system may dictate pressures greater than 100 psi in the pipelines at the source.

Transient responses to a change in system status, or surges, often result in significant increases in system pressures. It is possible to perform computer analyses of these transients to determine the total static and dynamic pressures which will occur throughout the system. Typically pressure transients of 100 psi, or greater, can occur in water systems. (Refer to Section 4 for greater detail.)

High transient pipeline pressures will also occur in

pump stations when control valves open and close while the pump is in operation. The resulting high "shut-off head" will exceed normal system pressures as the flow from the pump approaches 0 gpm.

5.3.2 External Loads

External loads include soil and live loads.

5.3.2.1 Soil Loads

Soil loads which generate stresses in the pipe include the trench earth load, and in certain areas of the country can include horizontal or vertical loads opposing the trench earth load, or both, due to frozen soils or clay soils which have the potential to expand or contract when changes in moisture content of the soil occurs. Stresses in the pipe caused by expansion or contraction of the soils can be significant and can require an increase in the thickness or strength of the pipe material.

5.3.2.2 Truck Loads

Truck loads must be considered if the pipeline is buried under streets or driveways or if future traffic is anticipated. Because traffic loads are dynamic, an impact factor is applied to increase the truck loading. Depth of cover over a pipe will influence the amount of truck load transmitted to it. Extra consideration should be given to loads that are transmitted to the pipe by heavy construction equipment or pipe with very little earth cover.

5.4 Thrust Restraint

Thrust forces, the resultant of internal pressure, occur at any point in a piping system where the direction or cross sectional area of the pipeline changes. Therefore, thrust forces occur at bends, reducers, tees, deadends and closed or partially closed valves. Pipe joints and fittings are most often mechanical or push-on joints. Neither of these joints provide restraint to counteract thrust forces that, unless restrained externally, can result in separation of pipe joints, pipeline or fitting breakage and significant leakage of water and loss of system pressure.

Pipeline thrusts are vector forces. For example, if a pipeline bend and reducer are placed in series, the thrust angle and force are significantly different than the thrust force that occurs solely from the bend.

For buried pipelines, thrust restraint is achieved

by transferring the thrust force to the soil structure surrounding the pipeline. The objective of the design is to distribute the forces to the soil structure in such a manner as to eliminate excessive deformation of the soil which allows movements of the pipeline components and separation of pipe joints. In evaluating soil deformation, bearing capacity, elastic deformation and creep and soil-pipeline friction must be considered. Thrust restraint is typically provided by concrete thrust blocks at fittings, concrete thrust walls along the pipe, or mechanically restrained pipe joints. In restrained joint design, thrust forces are contained within the piping system by utilizing mechanical connections to transmit tension across a bell and spigot joint or using a welded joint. Such systems must consider expansion and removal of valves and other pipeline appurtenances.

5.4.1 Concrete Thrust Blocks

5.4.1.1 General

By using a concrete thrust block placed between undisturbed soil and a pipeline bend or fitting, resistance to pipe joint separation is provided by developing the passive resistance of the undisturbed soil beyond the pipeline trench. Full passive resistance is only obtained at relatively large translations. To limit translation to acceptable values and protect the integrity of the pipe joint, conservative values of passive earth resistance, and a factor of safety of 3 or more, are recommended. If the pipeline is constructed in a fill or backfill adjacent to a structure, or below the groundwater table, it is recommended that the pipeline be restrained by means other than concrete thrust blocks. Generally the compaction of fill and backfill soils is not uniform or consistent, resulting in unpredictable and low passive pressures. Also, passive soil resistance will vary for concrete thrust blocks constructed against sloping ground surfaces. Therefore, when a pipeline is constructed along a ground surface which slopes downward from the pipeline grade, it is also recommended that the pipeline be restrained by means other than concrete thrust blocks.

5.4.1.2 Design

Various design aids and nomographs are available to provide general guidelines and recommendations for the design of thrust blocks. These design aids and nomographs make numerous assumptions about the soil and trench conditions. It is recommended however, that a geotechnical investigation make site specific recommendations for the design of concrete thrust blocks. With design parameters provided from the investigation along with the thrust force, concrete thrust blocks may be sized by acceptable theoretical analysis. Caution should be excercised to use only the lateral bearing value of an unconfined wedge of soil, rather than the vertical load bearing value for confined soils. Computer analyses of the block configuration using two-dimensional beam on elastic foundation or three-dimensional finite element computer programs have more recently been used to calculate bearing face area and soil displacements due to the thrust from the concrete block. Thrust block displacements should be limited to protect the integrity of the pipe joints. Since "stab depth", the length the spigot end of the pipe projects into the bell of the connecting pipe, varies with pipeline diameter and materials, acceptable values of displacement will also vary. If the pipeline joints are deflected due to alignment variations, the effective stab depth is also reduced.

5.4.1.3 Construction

Concrete thrust blocks should be carefully constructed in and against undisturbed soil. Generally a bond breaker (i.e. plastic wrap) is placed between the pipeline fitting and the thrust block to permit differential vertical displacements. The concrete thrust block should not be permitted to encase the pipe fitting and joints. Since thrust forces are vectors, the vertical face of the concrete thrust block which bears against the soil should be constructed perpendicular to the thrust force. Should the face not be perpendicular to the force, the passive forces developed will vary and partial failure of the soil and displacement of the block may occur. The bearing area of the thrust block must also be symmetrical about the resultant force vector to make the bearing pressure on the soil as uniform as possible. Eccentric loading can dramatically increase soil bearing pressures above those calculated for uniform distribution of pressure. If the soil exhibits any inelastic behavior, permanent displacement of the concrete thrust block may also occur. If displacement is the result of a pressure surge, repetitive surges could ultimately cause pipeline restraint failure, unless allowed for in the design.

5.4.2 Restrained Joint Pipelines

Pipe manufacturers have developed various types of interlocking pipe joints which resist separation of the pipe joint due to thrust. By using restrained joints, pipe to soil friction interaction and, in some cases passive soil pressures, can be developed to resist thrust. The objective in designing a restrained joint thrust restraint system is to determine the length of pipe necessary to develop soil friction resistance greater than the thrust.

Frequently ductile iron and steel pipelines have protective coatings to protect the pipeline from corrosive soils. These coatings can reduce the friction between the soil and the pipeline; thus the length of restrained joint pipe should be increased proportionately.

5.5 Valves

5.5.1 General

There are numerous types of isolation, control, and relief valves that are used in pipelines. The operational characteristics of the system, including pressure, pressure reductions, flow volume, and flow velocity, need to be considered in the selection of valves.

5.5.2 Isolation Valves

Gate, butterfly and plug type valves are usually used to isolate pipeline systems. Generally, gate valves are used for water pipelines smaller than 16-inches where throttling of the valve is not required for system operation. Butterfly valves are generally used in water distribution and transmission mains larger than 16 inches. Butterfly and plug valves are used in pumping facilities where modulation of the valve is used for flow control.

Valve spacing is a function of maintenance and operation considerations. In distribution systems, a general guideline is to locate isolation valves such that no more than 20 services would be without water should a break occur in a pipeline. In addition, it is desirable to locate valves so that no more than one fire hydrant would be out of service should a break occur. Valve spacing in transmission mains frequently requires consideration of future connections to the pipeline, control of the pipeline should a break occur and construction considerations, such as pressure testing or disinfection. Generally in lines 16-inches to 36-inches in diameter, valves are spaced 1/2 mile apart. For lines larger than 36-inches valve spacing is primarily a function of economics and system operation requirements.

The selection of the type of valve in distribution systems should also consider future maintenance and cleaning operations. Butterfly valve discs interfere with cleaning processes. Where cleaning is not expected this is not a problem. If cleaning is expected some gate, plug and ball valves are advantageous.

When selecting valves, it is important to specify an operation direction, open right or open left, which conforms with the utility standards.

5.5.3 Control Valves

Control valves include valves such as pressure reducing valves (PRV), pressure reducing-pressure sustaining valves (PRPSV), pump control, check, backflow prevention and altitude valves. Each of these valves has specific applications in piping systems. Selection of the various types of valves requires a thorough knowledge of desired system operation and valve operation, design and limitations.

Pressure reducing valves are required in multi-pressure zone systems where is desired to supply water from one zone to a lower pressure zone. Locating pressure reducing valves is critical considering the unidirectional flow of water from the high zone to the low pressure zone and the need to limit maximum pressures in the higher zone. Additional pipelines may be required in the pressure zone to maintain system pipeline service redundancy or "looping". Frequently it is necessary to place pressure reducing valves in parallel to permit operation of the system over a wide range of flow conditions. Valves placed in a series may be required when large pressure reductions are necessary.

Pressure reducing - pressure sustaining valves, not only reduce pressure from one zone to a lower zone, are designed to maintain a set pressure upstream of the valve. Without the pressure sustaining feature, low pressures can occur in the higher zone when demands in the lower zone downstream of the valve generate high flows through the valve.

Pump control valves are placed in pump and well facilities to control pump start-up pressure surges in the transmission or distribution pipeline systems. These valves are designed to open or close over a selected time period in order to gradually change system operational pressures. In addition these valves can be designed as check valves.

5.5.4 Relief Valves

Surge anticipating (surge relief) valves are often placed in booster pump and well facilities to provide relief from system surges. These valves open, venting the pressure surge from the system.

Air and vacuum relief combination valves are used in transmission pipelines at critical pipeline profile high points to release air and gas that may accumulated at those locations. They are designed to also admit air when a system failure occurs to prevent the development of a vacuum in the pipeline which may collapse a portion of the pipeline structure. Since valve operation may be infrequent two valves are sometimes placed are a single location to provide operation redundancy. Pipeline size and profile, along with flow, temperature and time considerations establish the capacity of the valve at a particular location.

5.6 Fire Hydrants

Fire hydrants are used to provide high volume flows for fire protection in water distribution pipelines and as blow-offs at critical low points in some transmission pipelines.

Fire hydrants should be located and specified in accordance with the requirements of the fire department providing fire protection for the area. Frequently fire departments will require isolation valves in distribution pipelines between hydrants in order to minimize service loss in the event of a distribution pipeline break. In addition, fire departments typically require "rolling" access to a fire hydrant. Fire hydrants at the end of a cul-de-sac are generally difficult to view and may not be easily accessible during a fire. However, hydrants located at a corner provide easy access and are highly visible. Fire hydrants usually have an isolation gate valve on the hydrant branch pipe for maintenance considerations.

When a fire hydrant is used as a blow-off on a transmission line, the main line tee is usually angled downward from the main pipeline to promote effective draining and cleaning of the line when the blow-off is used. Frequently, blow-off hydrants and pipeline isolation valves are required at crossings of streams and rivers where the profile of the pipeline is lowered under the river or stream. This enables the crossing to be isolated in the event of a break, and after repair flushing and disinfection.

5.7 Distribution Pipeline Blow-Offs

Generally fire hydrants are used to drain or flush pipelines in distribution systems. However, small diameter blow-offs are usually located at the end of cul-de-sacs and other dead ends, instead of fire hydrants.

5.8 Transmission Pipeline Blow-Offs

On large diameter transmission pipelines, blow-offs incorporate a gate valve in a manhole vault which provides a bubbler type drain for the main pipeline. Frequently blow-offs are not only located at critical low points in the pipeline profile but next to a manway pipeline access inlet.

5.9 Sewage Force Mains

Sewage force mains are pressure pipelines used to convey wastewater from lift station facilities. Wastewater has essentially the same physical properties affecting hydraulic design as water at the same temperature. Solids and grease in wastewater however, require special consideration in the design of the force main. The optimum size line requires the consideration of several factors, including velocity, quality of wastewater (size of solids and amount of grit), operating costs of the facility, ease of access for maintenance (rodding, pigging, flushing or jet cleaning) and present and ultimate pipeline flow volumes. Force mains should be designed to operate full to minimize the accumulation of hydrogen sulfide gas.

Since force mains are designed as pressure pipelines, restraint is required similar to water distribution and transmission pipelines, for both static and surge conditions.

Sewage force mains should be continuously sloped upward to a free discharge or else high points in the force main profile should have an automatic valve for venting the accumulation of air or other gases. Provision should be made for cleaning small diameter orifices of the valve.

5.9.1 Line Velocities

Force mains should be sized to minimize future maintenance requirements and provide self-cleaning flow velocities. The generally accepted minimum velocity is 2 fps. Velocities lower than 2 fps may result in the deposition of solids and accumulation of grease on the walls of the pipeline. The deposition of solids and accumulation of grease will affect the hydraulics and capacity of the force main, thus causing greater facility and main maintenance.

Velocities greater than 6 fps may cause excessive headloss and abrasion of pipeline walls if grit is present in the wastewater. However, velocities as high as 10 fps are not uncommon.

Designs with higher velocities should accomodate the greater surge potential by utilizing proper surge control devices and elimination of entrapped air.

5.9.2 Variable Facility Capacity

For lift station facilities having variable flows, consideration should be given to using multiple pumps and force mains or using variable frequency pump drives. The pipelines should be designed to meet minimum flow requirements over the facility flow range.

SECTION 6 - PIPE MATERIALS

6.1 General

Over the years, voluntary standards and manuals of practice which describe minimum levels of product quality and performance necessary for useful service in the conveyance of water and wastewater have been developed by committees within national organizations. These standards are not specifications, but can be made part of project contract documents by reference. They cover external load and internal pressure design, available sizes, quality of materials, installation practices, and information that relate to the hydraulics of flow, such as linings and actual inside dimensions.

Standards of the following organizations cover materials generally used for piping and accessories in North America:

American Waterworks Association	AWWA
American Society For Testing Materials	ASTM
American National Standards Institute	ANSI
Canadian Standards Assoc.	CSA

National standards committees are composed of consumer, producer, and general interest members.

In addition to the aforementioned standards and manuals of practice, other recommended standards and manuals of practice developed by manufacturers and manufacturers' associations are available for guidance in design and construction.

6.1.2 Permeation

The American Water Works Association is adding to their pipe standards for all pipe materials the following statement:

> The selection of materials is critical for water service and distribution piping in locations where there is likelihood the pipe will be exposed to significant concentrations of pollutants comprised of low molecular weight petroleum products or organic solvents or their vapors. Research has documented that pipe materials such as polyethylene, polybutylene, polyvinyl chloride and asbestos cement; and elastomers, such as used in

jointing gaskets and packing glands, may be subject to permeation by lower molecular weight organic solvents or petroleum products. If a water pipe must pass through such a contaminated area or an area subject to contamination, consult with pipe manufacturers regarding permeation of pipe walls, jointing materials, etc., before selecting materials for use in that area.

6.2 Cast Iron Pipe

Cast iron pressure pipe is no longer manufactured in North America and the associated standards have been withdrawn. Historical standards which cover existing cast iron pipe are available through AWWA and ANSI. Cast iron is now commonly referred to as gray cast iron.

6.3 Ductile Iron Pipe

The minimum requirements for manufacturing, testing and installation of ductile iron pipe and associated components are covered by the following standards:

ANSI/AWWA C104/A21.4	Cement-Mortar Lining for Ductile Iron Pipe and Fittings for Water
ANSI/AWWA C105/A21.5 & ASTM A674	Polyethylene Encasement for Ductile Iron Piping for Water and Other Liquids
ANSI/AWWA C110/A21.10	Ductile Iron and Gray Iron Fittings, 3-inch through 48-inch for Water and Other Liquids
ANSI/AWWA C111/A21.11	Rubber Gasket Joints for Ductile Iron and Gray Iron Pressure Pipe and Fittings
ANSI/AWWA C115/A21.15	Flanged Ductile Iron Pipe with Threaded Flanges
ANSI/AWWA C150/A21.50	Thickness Design of Ductile Iron Pipe
ANSI/AWWA C151/A21.51	Ductile Iron Pipe, Centrifugally Cast in Metal Molds for Water and Other Liquids
ANSI/AWWA C153/A21.53	Ductile Iron Compact Fittings, 3-inch through 16-inch, for Water and Other Liquids

ANSI/AWWA C600	Installation of Ductile Iron Water Mains and Their Appurtenances
ANSI/AWWA C606	Grooved and Shouldered Joints
ANSI/AWWA C651	Disinfecting Water Mains
ASME/ANSI B16.1	Cast Iron Pipe Flanges and Flanged Fittings
ASME/ANSI B16.42	Ductile Iron Flanges and Flanged Fittings

Ductile iron pipe is manufactured with a strength designation of 60-42-10, which corresponds to 60,000 psi minimum tensile strength, 42,000 psi minimum yield strength and 10% minimum elongation. Ductile iron as used in the manufacture of pressure pipe contains approximately 3.5% carbon by weight in the form of spheroids or nodules. This carbon content causes ductile iron pipe to have the same corrosion characteristics of gray iron, along with the strength and ductility much like that of steel.

Ductile iron pipe is designed as a flexible conduit. The pipe wall thickness is determined by separately considering internal pressure and external load. External load is based on two design considerations; limitation of ring bending stress and ring deflection. Earth load is computed by adding trench load and traffic load with impact.

The design internal pressure includes the working pressure plus 100 psi for water hammer allowance.

Manufactured in 3- through 64-inch nominal diameters, ductile iron pipe is normally cement mortar lined and seal coated to prevent internal corrosion and provide good flow characteristics. Other linings are available for conveyance of aggressive fluids such as septic sewage and industrial wastes.

The slip, or push-on, joint is most commonly used; however many joint types are available, including mechanical, flanged, grooved and shouldered, ball and socket river crossing, and restrained joints. All of these have some flexibility. A full line of gray iron and ductile iron fittings is available.

A maximum of 3% ring deflection is specified for cement lined pipe. Flexible linings are available such as polyethylene and epoxies which allow a 5% deflection to be considered.

Advantages of ductile iron pipe include its high strength against tensile, beam and impact loads, even at normal temperature extremes. It can be cut in the field and fitted together easily to meet changing field conditions. It can be direct tapped without saddles and most thrust restraint methods can be used. The actual inside diameters of ductile iron pipe are greater than nominal, resulting in greater flow capacity.

The disadvantage of ductile iron pipe is its susceptibility to corrosion in hostile environments unless suitable corrosion protection is provided. External corrosion protection is normally achieved by use of 8 mil thick, loose fitting polyethylene encasement.

6.4 Concrete Pressure Pipe

The minimum requirements for materials, manufacturing, testing, and installation of concrete pressure pipe are set forth in the following standards:

AWWA C300	Reinforced Concrete Pressure Pipe, Steel-Cylinder Type, for Water and Other Liquids
AWWA C301	Prestressed Concrete Pressure Pipe - Steel-Cylinder Type, for Water and Other Liquids
AWWA C302	Reinforced Concrete Pressure Pipe, Noncylinder Type, for Water and Other Liquids
AWWA C303	Reinforced Concrete Pressure Pipe, Steel Cylinder Type, Pretensioned, for Water and Other Liquids
AWWA M9	Concrete Pressure Pipe
ASTM C361	Reinforced Concrete Low-head Pressure Pipe
ASTM C648	Steel Wire, Hard Drawn for Prestressing Concrete Pipe

Concrete pressure pipe is designed as a rigid or semi-rigid conduit in accordance with one or more of the above standards. Typically, the manufacturer designs and produces pipe to meet specified internal pressure and external load requirements. A pipe laying schedule is required so that pipe, fittings, valves, etc., are placed at the correct location.

One or more of the several types of concrete pressure pipe is available in diameters ranging between 10 and 200 inches, but the most commonly furnished sizes are 16 through 96 inches. Compression type rubber gasket joints with confined solid O-ring gaskets are standard, with steel ring or concrete joint surfaces. External joint recesses are grouted to protect from corrosion the metal joint rings. Internal joint recesses may be protected by either cement, metalizing or other protective lining.

Restrained joint systems, including harnessed, welded, or special tied joints are available. Fittings are made either of modified pipes or of concrete coated and lined steel plate.

Advantages of concrete pressure pipe include shorter restraining distances due to its heavier weight. The protective lining is taken into account in design as an integral part of the structure. It is designed to specified service conditions. As a rigid conduit, it relies less on the trench bedding and backfill for support of external loads.

Disadvantages of concrete pressure pipe are that it is custom designed and changing field conditions require the design and manufacture of new fittings and closure pieces. Field cutting and tapping are more complex and it's heavier weight affects transportation costs.

The mortar coating protects the steel components from corrosion. However, recommendations are made that additional exterior corrosion protection methods should be used in adverse soil environments.

6.5 Steel Water Pipe

The applicable standards relating to materials, manufacture, testing, and installation of steel water pipe are:

AWWA C200	Steel Water Pipe 6 inches and Larger
AWWA C203	Coal-Tar Enamel Protective Coatings for Steel Water Pipe
AWWA C205	Cement-Mortar Protective Lining and Coating for Steel Water Pipe
AWWA C206	Field Welding of Steel Water Pipe Joints
AWWA C207	Steel Pipe Flanges
AWWA C208	Dimensions for Steel Water Pipe Fittings

AWWA C209	Cold Applied Tape Coating for the Exterior of Special Sections, Connections and Fittings for Steel Water Pipelines
AWWA C210	Liquid Epoxy Coating Systems for the Interior and Exterior of Steel Water Pipelines
AWWA C213	Fusion-bonded Epoxy Coating for the Interior and Exterior of Steel Water Pipelines
AWWA C214	Tape Coating Systems for the Exterior of Steel Water Pipelines
AWWA C215	Extruded Polyolefin Coatings for the Exterior of Steel Water Pipelines.
AWWA C602	Cement-Mortar Lining of Water Pipelines in Place
AWWA M11	Steel Pipe - A Guide for Design and Installation

A flexible conduit, steel water pipe is manufactured from various grades of materials. The yield point usually ranges upward from 30,000 psi. Wall thickness is based on internal pressure including surge, external loadings of earth and live loads, buckling pressures, and practical considerations of pipe handling and welding.

An allowable deflection for various lining and coating systems that are often accepted are:

Mortar lined and coated - 2% of pipe dia.
Mortar lined and flexible coated - 3% of pipe dia.
Flexible lining and coated - 5% of pipe dia.

Available in diameters 6-inches and larger, steel water pipe is most commonly furnished in 16- through 60-inch sizes. Numerous coatings and linings are available, including coal-tar enamel, epoxy, and cement mortar which provides good flow characteristics. Linings with Coal-Tar based material are not recommended for potable water service. External tape or other type coatings are also employed.

Welded, grooved or fabricated rubber gasket joints are most commonly used. With the exception of welded joints, they allow joint flexibility. Restrained joints are available for thrust control. The wide range of welding and fabrication processes applicable to steel pipe makes possible fittings of all types.

Steel pipe requires a pipe laying schedule so that fittings, valves, etc., will be constructed at the desired location. Any changes or changing field conditions require design and manufacture of new pieces. Fittings are made as required for particular projects.

External pipe joints are normally coated in the field with materials similar to those used on the main body of the pipe. On cement lined pipe, cement grouting the internal joint space is required.

Advantages of steel pipe are its adaptability to any configuration or design parameters. High strength steels make it practical for use under extreme operating conditions.

Disadvantages are related to the thin walls, which are subject to buckling or vacuum collapse unless designed to withstand these conditions. When steel pipe is coated or lined with concrete, the long lengths must be handled carefully to help preclude cracking of the mortar. Pipe walls are subject to corrosion unless suitable corrosion protection is provided. Cathodic protection usually supplements the external coating.

6.6 PVC Pressure Pipe

The following standards cover materials, manufacture, design, and installation of polyvinyl chloride pressure pipe:

AWWA C900	Polyvinyl Chloride (PVC) Pressure Pipe, 4 In. Through 12 In., for Water Distribution
AWWA C905	Polyvinyl Chloride (PVC) Water Transmission Pipe, Nominal Diameters 14 In. Through 36 In.
AWWA M23	PVC Pipe - Design and Installation
AWWA C907	PVC Pressure Fittings for Water 4 In. Through 8 In.
ASTM D1785	Standard Specification for PVC Plastic Pipe, Schedules 40, 80 and 120 (1/8" through 12" diameters)
ASTM D2241	Standard Specification for PVC Pressure Rated Pipe (SDR Series) (1/8" through 36" IPS O.D.)
CSA B 137.2	PVC Injection Moulded Gasketed Fittings for Pressure Application

CSA B 137.3 Rigid Poly (Vinyl Chloride) (PVC) Pipe for Pressure Application

PVC pipe is designed as a flexible conduit, the wall thickness being based on internal pressure. The pipe is then checked for adequacy in resisting external loads.

It is classed by design in terms of Standard Dimension Ratio ("SDR") which is the outside diameter of the pipe divided by the pipe wall thickness. Dimension Ratios ("DR") have been established which relate to a given pressure class.

PVC is a flexible material and is dependent on soil compaction around the pipe to maintain an allowable ring deflection. A maximum 5% ring deflection is recommended.

Manufactured in 1/8- through 36-inch nominal diameters, PVC pipe has a nonporous interior surface which results in good flow characteristics. No internal linings are required.

Compression type rubber-gasketed push-on joints are standard with PVC pipe manufactured to ASTM D2241, AWWA C900 and AWWA C905. A standard line of gasketed PVC, cast and ductile iron and steel fittings is available. Joint restraint devices manufactured specifically for PVC pipe are available.

Because PVC pipe is non-metallic, its greatest advantage is its immunity to nearly all types of external corrosion. Its light weight makes it easy to handle, and it is easy to field cut.

Disadvantages of PVC is it's vulnerability to ultraviolet rays when left exposed to direct sunlight for extended periods of time and is subject to fatigue failures when subjected to cyclic pressures. Being a thermoplastic, it has a high coefficient of expansion and contraction. Tensile and impact strength are affected also by temperature changes.

Tapping requires the proper use of cutting/tapping tools specifically manufactured for PVC pipe.

It is difficult to locate without tracer wire. Tracing wire can be buried with pvc pipe to facilitate future location.

6.7 Fiberglass Reinforced Thermosetting Resin Pipe

This pipe is commonly referred to as fiberglass pipe. The minimum requirements for materials, manufacturing, design, testing and installation of fiberglass pipe are set forth in the following standards:

AWWA C950	Fiberglass Pressure Pipe
ASTM D3754	Fiberglass Pressure Sewer & Industrial Pipe
ASTM D3839	Buried Installation for Fiberglass Pipe
ASTM D3840	Fiberglass Pipe Fittings
ASTM D4024	Fiberglass Flanges
ASTM D4161	Fiberglass Pipe Gasket-Sealed Joints

Fiberglass pipes are manufactured in diameters from 1" to 144". They are produced by both filament winding (fixed I.D.) and centrifugal casting (fixed O.D.) processes. Fiberglass pipes are constructed using glass fibers to reinforce thermosetting resins. Fillers and aggregates may be included to meet the design requirements. Corrosion resistance is inherent to the materials so secondary linings or coatings are not needed.

Fiberglass pipe is designed as a flexible conduit with a typical allowable deflection limit of 5%. The required pipe wall thickness is determined by separately considering internal pressure and external load. Pipe is designed for adequacy in buckling taking into consideration external loads, vacuum pressure, ground water level above the pipe and pipe stiffness. Pipe is also checked for adequacy in combined loading (external loads and internal pressure).

A wide range of joining systems are used including bell and spigot, gasketed fiberglass couplings, flanges, threaded, butt and overwrap and mechanical couplings. Standard and special fittings are available manufactured by compression molding, filament winding, contact molding or mitering of pipe sections. Fiberglass pipe can also be used with steel or ductile iron fittings.

Advantages of fiberglass pipe include its inherent corrosion resistance, lightweight construction, fabrication versatility, and good hydraulic characteristics.

A disadvantage of fiberglass pipe is that it requires bedding and backfill requirements that are more severe than for rigid pipe. Some types of fiberglass pipe are susceptible to localized deterioration where they are field cut unless suitable protection is provided. For future location tracing wire is needed. Tapping procedures are different from those of ferrous materials.

6.8 Polyolefin Pipe

The applicable standards for materials, manufacture, design and installation of these piping materials are as follows:

AWWA C901	Polyethylene Pressure Pipe, Tubing and Fittings, ½ in. through 3 in., For Water
AWWA C902	Polybutylene Pressure Pipe, Tubing and Fittings, ½ in. through 3 in., For Water
ASTM D2666	Polybutylene (PB) Plastic Tubing
ASTM D2774	Underground Installation of Thermoplastic Pressure Piping
ASTM F645	Selection, Design and Installation of Thermoplastic Water Pressure Piping Systems
ASTM F809	Large Diameter Polybutylene Plastic Pipe

Polyethylene pressure tubing and piping is available in sizes from ½ inch through 63 inch diameters while polybutylene pressure tubing and piping is available in sizes from ½ inch through 42 inch diameter.

Polyolefins (polyethylene and polybutylene) are designed as flexible conduits. The pipe wall thickness is determined by considering internal pressures and external load as separate entities. Considerations for external pressures include ring deflection, possible hydrostatic loadings, trench backfill as well as any traffic loadings. Internal pressure considerations include working pressure as well as allowance for water hammer. The combination of working pressure plus recurring surge pressure should not exceed 1.5 times the pressure class selected.

Joining pipe sections is done by a process known as thermal butt fusion. This process joins the pipe through a combination of heat and pressure.

High density polyethylene pipe typically range between 4 and 48 in. nominal diameters. Joints are fused butt welds. It is frequently used as a liner as it conforms to the contours of existing pipe inside diameters.

Tapping procedures are different than for ferrous materials.

The advantages of these materials are they are

lightweight, they are easily field cut and inert to most forms of corrosion.

A disadvantage of polyolefin pipe is that they require better bedding and backfill requirements than more rigid pipe. Thicker walls for greater loads reduce the inside diameter. They have a high coefficient of expansion and tracing wire is needed for future locations.

6.9 Asbestos-Cement Water Pipe

The standards applicable to the materials, manufacture, design, and installation of asbestos-cement pressure pipe are as follows:

AWWA C400	Asbestos-Cement Distribution Pipe, 4 In. Through 16 In. for Water and Other Liquids
AWWA C401	The Selection of Asbestos-Cement Distribution Pipe, 4 In. Through 16 In., for Water and Other Liquids
AWWA C402	Asbestos-Cement Transmission Pipe, 18 In. Through 42 In., for Potable Water and Other Liquids
AWWA C403	The Selection of Asbestos-Cement Transmission and Feeder Main Pipe, Sizes 18 In. Through 42 In.

Asbestos-cement pipe is used in pressure piping applications. Design is on the basis of a rigid conduit, considering combinations of internal and external loadings together to determine the required wall thickness.

Joints are made with gasketed couplings, and adaptors are available for joining other pipe materials. Cast or ductile iron fittings are normally used.

The advantages of asbestos-cement pipe is its low cost, relative light weight, and its composition which eliminates many concerns about corrosion. However, it has been found that asbestos-cement pipe does deteriorate when in contact with hostile environments and aggressive waters.

Disadvantages include its susceptibility to failure when loaded as a beam and upon impact. It is also difficult to locate using traditional methods.

In 1991, the Occupational Safety and Health Act became effective which gave a standard for exposure to Asbestos dust. Specific work practice methods are recommended for cutting, machining and tapping asbestos-cement pipe and for waste disposal.

6.10 Pipe Linings

6.10.1 Cement Mortar

Metallic pipes which have been cement lined show an excellent record in maintaining high flow coefficients as this lining is effective in preventing corrosion and tuberculation in the interior of the water pipe. The linings are placed centrifugally, which produces a dense mortar with a smooth surface. Concrete pipe standards allow either centrifugal or static placing of the cement lining.

Older existing water pipelines that are tuberculated inside may be rehabilitated in-place by cleaning, and then applying a cement mortar lining. Standards list the recommended thickness of lining for new and rehabilitated pipes.

6.10.2 Coal Tar Enamel

This type of lining has been used many years. AWWA C203 Standard covers application and material requirements for steel pipe. This standard is presently being revised to eliminate coal tar as a lining for potable water due to consideration of leaching contaminants into the water. Coal tar is available as a lining for ductile iron pipe for sewer service only.

6.10.3 Liquid Epoxy

This lining can be used for water or sewer service. In AWWA C210, it consists of one coat of a two-part chemically cured epoxy primer and one or more coats of a two-part chemically cured finish coat.

6.10.4 Fusion Bonded Epoxy

For steel pipe, this is a heat activated, chemically cured lining.

6.10.5 Polyethylene

Ductile iron pipe with a polyethylene lining is available in all sizes. It was developed for use in septic sewer or industrial waste service. It has a high flow coefficient and provides protection to the substrate, with no taste or odor imparted to the liquid.

6.11 Inside Pipe Diameter

The actual inside diameter of a pipe is an important factor to the hydraulic flow, as the capacity of the pipe is proportional to the 2.63 power of the inside diameter.

(Hazen-Williams formula)

$$Qgpm = 0.006756\ C\ D^{2.63}\ S^{0.54}$$

Thus, the I.D. has a more pronounced effect on flow capacity than does the C factor.

For transmission mains, the actual inside diameter of the pipe should be used in calculations for flow and pumping costs.

Ductile iron and centrifugally cast fiberglass pipes have standardized on the outside barrel dimensions. In all normally specified pipe classes and all sizes, these pipes have an internal diameter that is larger than nominal.

Steel water pipe is manufactured in a variety of outside and inside diameters. The designer should become familiar with the standard size so that actual internal diameters with allowances for lining thickness be used in hydraulic analyses.

Concrete pressure pipe is manufactured to nominal inside diameters.

Plastic pipe inside diameter varies with pipe diameter, DR (wall thickness) and whether the outside diameter is iron pipe size or cast iron pipe size. In the DR's usually specified, the I.D. is smaller than nominal.

6.12 Sewage Force Mains

Sewage force mains may be constructed of ductile iron, steel, plastic, fiberglass or concrete pressure pipe. Hours of down time with partially filled lines or high points in the line need careful consideration. Whenever air pockets or partially empty pipes are encountered, the potential for internal corrosion is increased. This potential can be controlled by maintaining a full flow or completely empty condition. For metallic or cementitious pipes, cement lining will suffice for corrosion protection if the line is always full. Other linings, such as polyethylene, epoxies and coal tar epoxies have proven effective along with the use of plastic or fiberglass pipes.

REFERENCES

1. Energy Savings with Ductile Iron Pipe, DIPRA, 1986.
2. Unibell PVC Pipe Handbook, 2nd Edition
3. Steel Pipe - A Guide for Design and Installation, AWWA M-11.
4. Guideline for the Protection of Concrete Cylinder Pipe in Adverse Environments, American Concrete Pressure Pipe Assoc.
5. Prestressed Concrete Pressure Pipe, AWWA C301-84.

SECTION 7 - PIPELINE ECONOMICS

7.1 General

The common economic comparisons used with the selection of pipe material, sizes, and other design features are described herein. The simple illustrations used are intended to explain the individual points to be considered. The combined effect of the several interrelated factors normally encountered is discussed. However, actual applications of these principals are usually complicated by many factors, so that a so-called "systems" approach, described in section 7.5, is often needed to obtain an overall economic solution. Also, the use of the term "pipeline" refers only to the pipeline, waterway, earthwork, and backfill designs and any appurtenances that are required for the ALTERNATIVE DESIGNS comparison.

7.2 Pipelines

7.2.1 Pipe Materials

Pipeline economic studies begin with the selection of the pipe material which will meet the selected design criteria. This involves achieving a balance between pipeline cost (including installation) and life expectancy of the pipeline, considering pressure, cover, and installation requirements suitable for the conditions. Many types of pipe material are available. Among these are steel, concrete, ductile iron, polyvinyl chloride (PVC), fiberglass, and polyethylene (PE). The type of material chosen for economic alternatives will be based on pressure requirements, suitability for installation conditions, longterm inherent strength and corrosion requirements. The service life of a pipe material varies widely, based on the potential for corrosion. Materials such as fiberglass, PE, or PVC do not react as readily to corrosive environments as concrete and steel but may lack long term strength requirements. Corrosion protection can be provided.

A simplified trade-off between life expectancy and cost is illustrated by the present worth analysis presented in Table 6. To evaluate pipe materials with life expectancies of 25, 50, and 100 years, consideration must be given to the replacement of the pipe with the shorter life so that each alternative design is compared over the same time span.

Table 6 also indicates that the pipe with the 25 year life has to be replaced at the end of 25, 50, and 75 years while the pipe with the 50 year life has to be replaced only at the end of 50 years. Therefore, the three alternative designs are evaluated on the basis of present worth of providing the same pipeline service over a 100 year period.

TABLE 6 - ECONOMIC COMPARISON OF PIPES WITH DIFFERENT USEFUL LIVES

Item	Alternative Designs		
	Pipe A	Pipe B	Pipe C
Estimated useful life, years	25	50	100
Initial cost per foot, installed	$35	$40	$45
Present Worth*			
Initial installation cost	$35.00	$40.00	$45.00
25 year replacement cost	8.15	-	-
50 year replacement cost	1.90	2.15	-
75 year replacement cost	0.45	-	-
	$45.50	$42.15	$45.00

*Based on interest rate of 6%

7.2.2 Pipe Size

Pipe is available in a large number of sizes between 4-inch and 36-inch based on the type of pipe being furnished. Consult industry standards and manufacturer's specifications for exact standard sizes of the type of pipe furnished. Standard sizes between 36-inch and 54-inch increase in 3-inch or 6-inch increments and for larger than 54-inch the pipe diameters vary, but generally increase in 6-inch increments. Specifying a pipe diameter other than a standard size for a given type of pipe is usually not economical, even though the pipe diameter chosen may more closely meet the computed hydraulic requirements. However, the substitution of a larger size for a specified smaller or less common diameter should be allowed as this will give more suppliers the opportunity to bid on a pipeline and this may reduce the overall cost.

Table 7 illustrates the relationship between pipe sizes and pipeline capacity, as well as the relationship between pipeline capacity and pipeline cost. For a given pressure loss, larger diameter pipes deliver larger flow rates. This table can be expanded for larger diameters. However, when selecting the pipe diameter, the effects of flow velocity must be taken under consideration. For velocities less than 1 fps transient effects of valve closures, pump shutdowns, etc., generally have little effect on the design pressure class of pipelines. When the velocity increases above 1 fps, the transient effects need to be evaluated and a design made which includes the proper pressure classes. A rule of thumb for pumped systems is that a velocity of between 5 and 7 fps is a good balance between pumping costs and pipeline diameter costs.

TABLE 7

Pipe Size (inches)	Velocity* (fps)	Capacity* (mgd)	Cost** ($)	Cost/ft/mgd ($LF)
4	1.8	0.1	19.20	192.00
6	2.3	0.3	25.42	84.73
8	2.8	0.6	27.65	46.08
12	3.6	1.8	40.74	22.63
16	4.3	3.9	50.11	12.85
20	5.0	7.0	75.27	10.75
24	5.5	11.4	101.75	8.93
30	6.5	20.2	122.58	6.07
36	7.3	32.6	157.99	4.85
42	8.0	48.5	200.01	4.12
48	8.6	70.0	217.24	3.10

*Based on head loss of 4 ft/1000 ft using the Hazen-Williams Formula and $C_H = 130$.

**Based on typical pipeline costs from 1987 Dodge Heavy Construction Cost Data, cover is 5 feet. (Diameter ≤ 24 inches. PVC class 150.) (Diameters > 24 inches, steel pipe, 3/8-inch wall.)

The cost per mgd of capacity decreases markedly as the pipe size increases. The last column in Table 7 shows that the cost per mgd of a 6-inch pipe is about 4 times the cost of providing a 1 mgd unit of capacity in a 24-inch pipe and 18 times the cost of the same capacity in a 36-inch pipe.

7.2.3 Economic Decisions

One common economic decision is whether to construct a larger diameter pipeline initially or put in a smaller pipeline first and construct a parallel pipeline at a later time. The example shown in Figure 25 indicates that for the conditions given, construction of a single 42-inch pipeline, would be economical only if the demand increases to more that 50 percent of the initial capacity within 13 years of the initial construction. However, if this additional capacity is not expected to be needed for 13 years, construction of a 30-inch pipeline at a later date would be more economical.

This analysis is based on a constant valued dollar. Inflation following initial construction will increase the dollar cost of future construction, but this increase will be moderated by cheaper dollars that will be easier to obtain because such cheaper dollars can be collected from the higher water rates which will follow the inflationary trend.

Conversely, a period of deflation following the initial construction would result in a lower dollar cost for a second pipeline, but such dollars would be correspondingly harder to raise during such a period.

7.3 Pumped Pipelines

7.3.1 Pumping Pressures Versus Pipe Wall Thickness

In small diameter pumped lines, higher pressures are often used to increase the length between pump stations. The minimum practical wall thickness of metal pipe often permits higher pressure operation. Therefore, the use of metal pipe with higher pressures and fewer pump stations is sometimes economical. In steel pipelines, greater than 30 inches in diameter, this may no longer be valid, because increased stresses for normal pressures require at least the minimum pipe wall thickness. The following tabulation shows the allowable internal pressure for the minimum wall thickness for each diameter. Factors other than handling or internal pressure, such as depth of burial or live loads, may require thicknesses considerably greater than those tabulated.

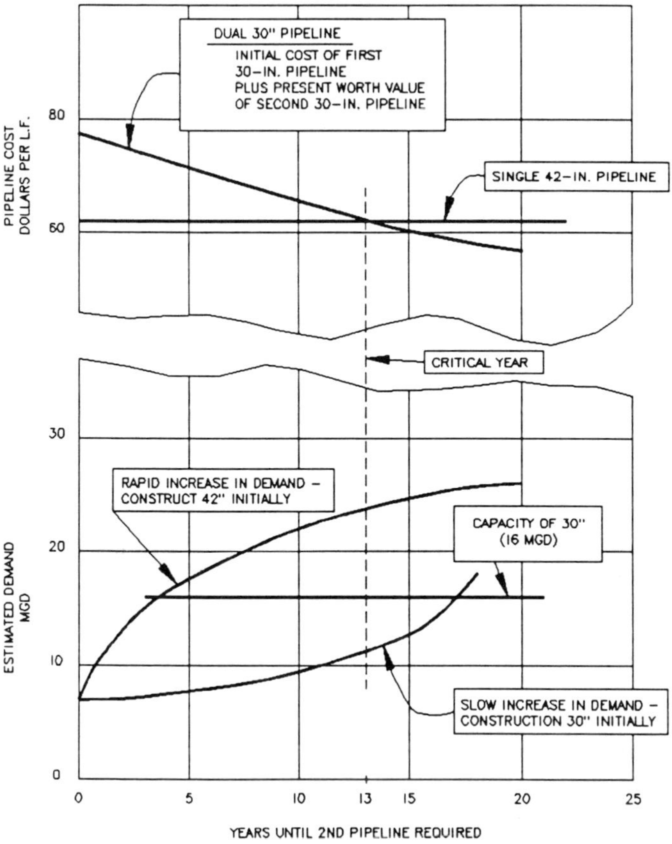

Figure 25
Economics of Prebuilding for Future Needs

ALLOWABLE PRESSURE
BASED ON "MINIMUM" PIPE WALL THICKNESS

Size	Minimum Wall Thickness (inches)			Allowable Pressure (psi)		
	Steel	Ductile Iron	PVC[1]	Steel[2]	Ductile Iron[3]	Polyvinyl Chloride (PVC)[4]
8	.0747	.25	DR25	280	350	100
12	.0747	.28	DR25	187	350	100
16	.0747	.30	DR41	140	250	100
20	.0747	.33	DR41	112	250	100
24	.1046	.33	DR41	131	200	100
30	.1046	.34	DR41	105	150	100
36	.1046	.38	DR41	87	150	100
42	.1046	.41	----	75	150	---
48	.2500	.46	----	156	150	---
54	.2500	.51	----	139	150	---
60	.2500	.54	----	125	150	---
64		.56	----		150	---

[1]DR - Dimension Ratio: Ratio of diameter to pipe thickness.

[2]Based on steel pipe using design of 15,000 psi and handling criteria in AWWA Steel Pipe Manual M-11 (1989)

[3]Based on ductile iron pipe with stress of 21,000 psi at net design thickness: per ANSI/AWWA Ductile Pipe Standard C150/A21.50 (1992). An allowance of 100 psi for water hammer over and above these pressures is included in the design.

[4]Allowable pressures based on AWWA C900 and AWWA C905. AWWA C905 has no allowance for water hammer.

Based on Table 8 and the indicated costs for pipeline construction, pumping station and power costs for either 30- or 36-inch pipe would produce about the same annual cost for continuous pumping at pipeline capacity. The 30-inch pipe is more sensitive to power costs, so that lower power costs would make it more favorable and would have the advantage of reduced first cost, because the pipe and installation costs would be lower. The 24-inch size would require either higher pressure pipe and pumping equipment or more than one pumping station. A very substantial reduction in power cost would be necessary to make the use of 24-inch pipe attractive. On the other hand, 42-inch pipe would have to be 20 percent cheaper than indicated with relation to the smaller sizes before it could be consider competitive. On Figure 26, the graphical solution shows 33 inches as the most economical diameter.

7.3.2 Pipe Size Versus Pumping Cost

One common analysis is the selection of an economical pipe diameter in a pumped system. The most economical pipe diameter is determined by balancing pumping costs against pipeline costs for various sizes of pipe. In a pipeline where gravity flow can be used economically, pipe diameter is selected to produce a friction loss that will just utilize the available head. For a pipeline in a pumped system consideration must be given to the distance between the intake and the terminal reservoir and the available head between these two locations.

The pipe size selection is based on an analysis of the costs involved in Table 8 and shown on Figure 26.

TABLE 8 - PIPE SIZE COMPARISON FOR PUMPED TRANSMISSION PIPELINES

Required Capacity - 16 mgd $C_h = 130$

Length - 20 miles (105,600 ft) Power @ $0.05/kw-hr

Compare Costs for 24-inch, 30-inch, 36-inch, and 42-inch pipes

	24-inch	30-inch	36-inch	42-inch
Cost/ft ($)	101.75	122.58	157.99	200.01
Total Cost ($)	10,744,800	12,994,448	16,683,744	21,121,056
H_L/100 (ft)	7.5	2.5	1.05	0.49
Total H_L (ft)	789.7	264.0	110.9	51.7
HP (100% Eff)	2222	739	310	145
Assume Eff.	80%	82%	82%	81%
Required HP	2778	900	378	180
Kilowatt	2072	671	281	134
Kw-hr/yr	18.2×10^6	5.87×10^6	2.46×10^6	1.17×10^6
Cost of power ($/yr)	910,000	293,500	123,000	58,500
Cost of Pumping Station ($)	2,734,000	1,072,000	679,000	495,000
Annual cost of Pump Station over 20 years (0.0871) ($)	238,131	93,371	59,140	43,115
Total annual costs of pumping ($/yr)	1,148,131	386,871	182,140	101,615
Annual cost of pipeline over 40 years (0.0664) ($)	713,454	859,511	1,107,800	1,402,438
Total annual costs ($)	1,816,585	1,246,382	1,289,940	1,505,553

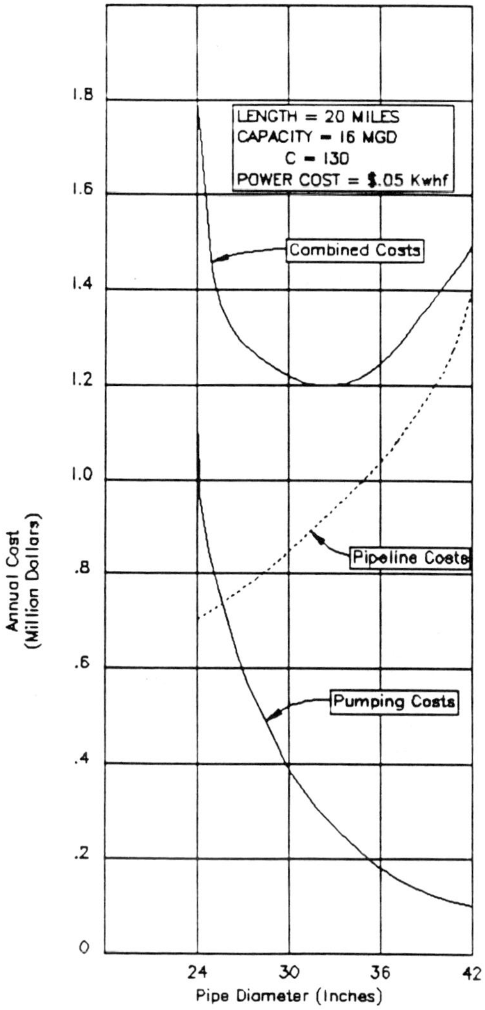

Figure 26
Economic Pipe Size for Pumped
Transmission Main at Full Capacity

The foregoing analysis is appropriate where a fixed quantity is to be pumped on a continuous basis beginning immediately upon construction of the pipeline. In actual practice these are seldom the conditions for water main construction. A water transmission pipeline is usually designed for a capacity to be required at some future date, so that during the initial years considerably smaller quantities are actually pumped through the line. As already illustrated, to design for a period of more than 15 or 20 years, depending on the rate of growth of the community being served may not be economical.

In design, actual internal diameters should be used when computing headloss and pumping costs, as the actual inside diameter can vary between pipe materials (see page 88).

Another factor which must be considered in water transmission pipeline design is the fact that the pipeline capacity must handle flow on the maximum day of the year which is substantially greater than the average flow for the year. In fact in many localities the maximum day flow may be as much as twice the average day flow.

These factors combine to substantially reduce the actual power costs from those shown in Figure 26. This is illustrated in Figure 27, where these factors have been considered and power cost reduced accordingly. As a result annual costs are reduced and the lower power cost for smaller size pipe may justify the selection of a size smaller than for a constant capacity line. In this example the size would be the same for either case, but in Figure 27, the 30-inch size is seen to be clearly more economical than 36-inch, particularly during the first 15 years of pipeline use.

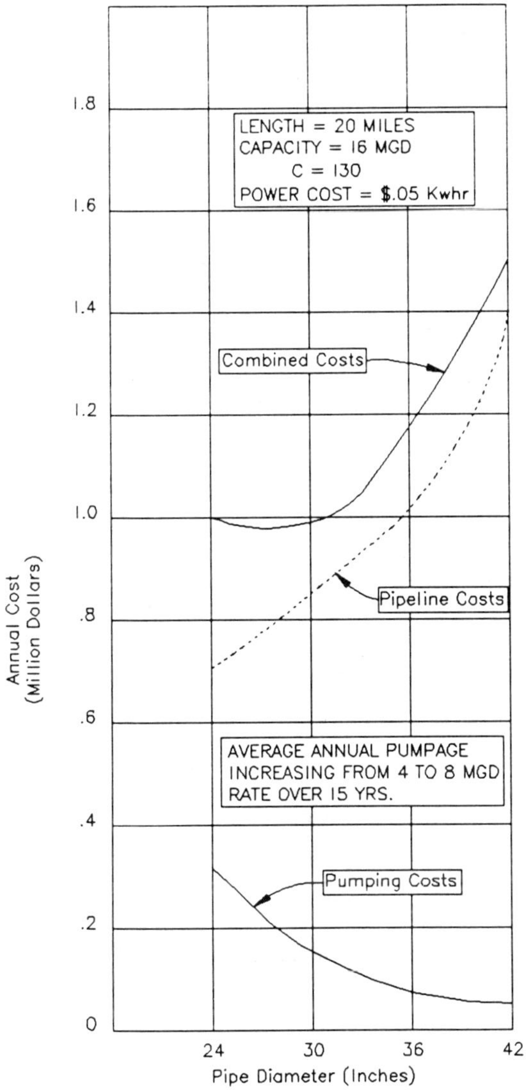

Figure 27
Economic Pipe Size for Pumped
Transmission Main at Partial Capacity

The pumping costs used in the above illustrations are based on a unit cost of power which is dependent only on the total quantity used. Electric power rates normally include several variables such as a "demand course" based on the maximum power demand rate over a given period of time (usually per year) which is independent of the unit cost of power used. The introduction of these factors would exaggerate slightly the pumping costs in Figure 26 and even more in Figure 27. In an actual analysis the introduction of the variable charge will tend to reduce the advantage of the smaller sized pipes by increasing power costs more rapidly, and so should be carefully considered in a refined analysis.

7.3.3 Increased Capacity

The designer is often faced with another economic decision to provide additional capacity in an existing system.

Consider a system where an existing intake pumping station discharges 16 mgd against 125 psi into a 30-inch pipeline which transmits water 17 miles and discharges into a terminal reservoir at 10 psi. The requirement is to increase the system capacity from 16 to 23 mgd. To provide this would require construction of 90,000 feet of a 24-inch parallel pipeline, or the construction of an intermediate pumping station in the existing pipeline which would have a capacity of 23 mgd against 270-foot head. Table 9 shows the analysis and the clear indications that the pumping station would be the most economical alternate under the conditions set forth.

Generally, providing intermediate pumping stations for larger pipelines, rather than higher pressure pipe, are more economical. Since total pumping head would be the same for each alternative considered, the annual cost of electric power would be comparable. However, the cost of higher pressure pipe for part of the pipeline would increase the pipe cost roughly in proportion to the increased pressure requirement, and would probably more than offset the increased construction cost for pumping facilities.

TABLE 9 - ECONOMICS OF INCREASING CAPACITY OF EXISTING SYSTEM

Existing Conditions

Intake pump station	Capacity	16 mgd
	Discharge Pressure	125 psi
Pipeline	Size	30 inch
	Length	17 miles
	Capacity	16 mgd

Requirements

Increase capacity to 23 mgd

Alternates

1. Construction of 90,000 feet of 24-inch pipeline Useful life 50 years.
2. Intermediate pumping station with capacity of 23 mgd @ 270 ft. of total head. Useful life 25 years.

Analysis

1. Initial cost of 24-inch pipeline	$9,157,500
2. Initial cost of pumping station	$1,123,000
Present worth of replacement of pumping station after 25 years ($1,123.000 X .233)	**261,659**
Present worth** of annual power cost over 50 year period	**$3,328,272**
Capitalized Cost for Pumping	**$4,712,932**

*Based on 6% interest rate.

**Based on (16 + 23)/2 mgd for first 20 years then 23 mgd for next 30 years

Power cost $0.05 per kw-hr. Efficiency 80%.

7.4 Storage

Pipelines are often designed to meet peak rates of demand. In many cases, the most economical solution is to provide storage to meet these peaks and, therefore, reduce the pipeline size between the source of supply and storage facilities.

In a simple system, let us assume a peak hourly demand of 32 mgd and a maximum one-day demand of 16 mgd. If a tank was economical, a 4 million gallon reservoir would be recommended (based on 20 percent of maximum day demand plus reserve) to provide storage.

Without storage, pumping facilities and the transmission pipeline would be sized to deliver 32 mgd. Under these conditions, a 42-inch pipeline would be required, whereas if 4 million gallons of storage were provided, a smaller pipeline (30-inch) would be adequate.

The economic evaluation of using storage reservoirs can be easily made for any given condition. However, the length between the source of supply and the proposed storage location is often the deciding factor. Therefore, a simple algebraic operation will determine the maximum pipeline length that provides the most economical pipeline and storage facility. In this analysis, $123 and $200 have been selected as the respective costs per foot of construction for a 30- and 42-inch pipeline, and $1,000,000 as the cost of constructing a 4 million gallon storage facility.

Solving for L, in feet:

$$123L + 1{,}000{,}000 = 200$$

$$323L = 1{,}000{,}000$$

$L = 3{,}000$ feet (Maximum economical length for larger pipeline)

Pumping facilities are normally a part of the typical system considered by the designer. When storage is provided, smaller pumping facilities can be installed at lower initial costs and lower annual operating expense. Consideration of pumping facilities in the economic analysis presented above would reduce the maximum pipeline length necessary to make the alternate with storage and smaller pipe preferable.

7.5 System Approach to Evaluation

An approach which many have used in the past has in recent years been formalized and popularized as the "systems approach." Using this approach the designer should include in his evaluation all of the factors pertaining to the optimum solution to his problem. Some of the factors most often involved have been presented above as separate problems.

The systems approach is essential if the optimum solution is to be made at each stage of system development. Often such analysis might indicate that a large pipeline capable of handling initial peak hour flows be installed first followed by the addition of pumping and/or storage facilities to provide for increased capacity and operational flexibility as additional capacity is needed over the years.

A true systems approach would, however not be confined to the transmission pipeline and related facilities but would include consideration of the inter-relationship between transmission, supply, treatment, and distribution facilities. Such an approach is needed to achieve the most economical overall design for major waterworks projects.

Computerized cost comparisons are common in which a very large number of variables, such as pipe diameter, flow rates, deliveries, financial costs, etc., may be assumed and estimates calculated for an optimal solution of the economic problems.

7.6 Summary

1. Pipelines having different useful lives may be evaluated over the life span of the most durable pipe by considering the present worth of replacement cost of the pipes having shorter useful lives.

2. Larger pipelines deliver large quantities of water more economically than smaller lines.

3. The decision to build a pipeline larger than is needed initially is economically justified when the future demand is expected to exceed the capacity of a smaller line before the present worth value of two smaller pipelines of equivalent capacity becomes less than the cost of the larger pipeline.

4. The most economical pipe diameter for a pumped system is generally established by determining the combined construction and operating costs for pipeline and pumping facilities over a range of standard pipeline diameters, so as to indicate the diameter where combined costs are

minimized. Usually the effect on power costs of partial capacity required during early years should be considered.

5. The provision of storage is economically justified when the distance between supply and storage facilities exceeds a certain length. This critical distance is determined by the condition when the additional cost of the larger pipeline and pumping capacity required without storage exceeds the cost of providing storage with a smaller pipeline. By inclusion of storage and a constant pumping rate, smaller pumping facilities and pipeline are required to meet normal daily fluctuations in demand.

6. The systems approach should be adopted where a detailed analysis is warranted by the size of the project. This requires careful analysis of the inter-relation of the several factors discussed above, as well as the relation of pipeline problems to all other aspects of the project.

SECTION 8 - SEISMIC DESIGN

8.1 Introduction

This chapter covers seismic resistant design of pipelines. Earthquake hazards are discussed followed by earthquake damage mechanisms. Design analysis techniques are presented to assess earthquake affects and recommendations included to resist earthquakes.

Water and wastewater pipelines are a critical part of water and wastewater lifeline systems required to maintain life functions and assist in disaster recovery operations. Water systems provide fire protection, crucial following an earthquake, water for drinking, and food preparation, as well as for sanitary uses.

Failure of the water transmission and distribution system in the 1906 San Francisco earthquake resulted in the City burning. Following the 1985 Mexico City earthquake, 3.5 million people were without water. Failure of water lines in the 1989 Loma Prieta earthquake cut water service to hospitals in Santa Cruz (Ballantyne 1990), limited fire fighting capacity in some areas of San Francisco, and put small water systems out of operation for as long as five months.

The performance of underground pipelines during recent earthquakes has ranged from good to poor. In general, significant pipeline damage has resulted from earthquakes of magnitude six or greater as measured on the Richter scale. Where large permanent ground movements have been observed, pipelines in those areas have either been sheared, separated or crushed. Fortunately, these permanent ground failure areas are limited and usually only affect a small percentage of pipe exposed to the earthquake.

In general, pipeline performance appears to depend minimally on pipeline diameter. Statistically, pipelines 16 inches and larger in diameter appear to have lower failure rates than smaller lines. This difference is however small when compared to the effects caused by such factors as pipe material, joint type, and the level and type of seismic hazard.

The performance of underground water lines has been studied significantly since the 1971 San Fernando earthquake. That earthquake afforded an opportunity to assess the performance of a number of pipeline systems.

Some conclusions that can be drawn were:

1. Performance varies significantly on whether lines were subjected to fault rupture, liquefaction, landslide, or only ground shaking. Failure rates range from 5 repairs per 1000 feet in surface fault rupture zones to as little as 0.015 repairs per 1000 feet in areas of pure ground shaking.

2. Pipe joint type is an important factor in establishing failure rates. In ground shaking areas, the more flexible the joint, the better the performance. For example, joints using resilient gaskets experienced few failures. Joints in steel pipe using gas welding did not perform well because of poor quality or heat effects on adjoining material. As a result, many hairline cracks occurred.

3. Corrosion played an important role in accelerating failures. Corrosion also leads to increased break rates in the years following the event.

4. The angle and direction at which the pipeline crosses the fault has a significant effect on the amount of strain experienced by the pipe and thus the performance of the pipe.

These finding have been substantiated by more recent earthquakes, including the 1983 Coalinga earthquake, the 1985 Mexico City earthquake, and the 1987 Whittier Narrows earthquake.

The 1989 Loma Prieta earthquake caused more than 1000 water line failures, exceeding the pipeline damage in the 1971 San Fernando event. A smaller number of sewer line failures is expected. Preliminary analysis indicates that similar conclusions can be drawn from that event.

8.2 Seismic Hazards

Seismic hazards include tectonic and geotechnical manifestations that may impact engineered systems.

8.2.1 Liquefaction and Landslide

Liquefaction is the fluidization of cohesionless soil usually as a result of increased pore water pressure. It may occur when three conditions are met:

1. Soil is cohesionless (i.e. sand)

2. Soil is poorly consolidated (i.e. loose)

3. Groundwater table is high, usually less than 30 feet below grade (i.e. saturated)

These conditions are often met in alluvial deposits and in man-made fills not properly designed. Standard penetration field tests or cone penetration tests can help establish liquefaction and consolidation susceptibility.

Consolidation or settlement can result when conditions one and two are met but the ground water table is deep. Consolidation does not result in loss of bearing but causes relative displacement.

Liquefaction can result in loss of bearing strength and flotation of submerged structures. Liquefied material can also induce lateral spread of overburden.

8.2.2 Ground Motion

Ground motion is the transient, recoverable ground deformations during seismic shaking. Shear and compression waves (S-waves and P-waves) are more predominant in near field events where as Rayleigh and Love waves are more predominant in far field events. Estimates of the maximum acceleration, velocity, and displacement of the ground can be made from a seismic hazard analysis. The peak ground velocity is an important parameter in establishing vulnerability to seismic shaking.

Ground motion can be amplified locally by soft soils. This was demonstrated in the Loma Prieta earthquake where extensive damage resulted from ground motion amplification in the San Francisco Marina District and in Oakland along highway 880.

Ground motion is sometimes amplified at the edge of a large geologic basin, or at soil type interfaces. Particular precaution should be employed at these locations.

Pipelines can be damaged when they can not accommodate earthquake wave induced relative movement along a pipeline, across a pipe joint, or at a pipe/building interface.

Seismic shaking can also induce increased internal pressure in a pipeline due to surge and water hammer effects. The increased pressure can lead to longitudinal splitting of the pipe. Wang (1990) indicates that this was the predominant failure mode for pipelines in the

1987 Whittier Narrows earthquake.

8.2.3 Fault Crossing

Tectonic movement between plates occurs at faults. Faults may extend to the surface such as in California. Relative movement may occur between the two sides of the faults. In other areas, such as in the Pacific Northwest and in the Eastern United States, surface faulting may not be a problem.

8.3 Damage Mechanisms

8.3.1 Introduction

This section considers earthquake hazards and identifies the mechanisms from which pipeline damage can result.

8.3.2 Liquefaction and Landslide

Pipelines can be damaged in lateral spreads and from flotation. Similarly, pipelines can be damaged where relative movement occurs as a result of consolidation. An example is at the transition from a pile supported structure to a direct buried structure where settlement occurs in the non-pile supported segment.

Landslides can result from earthquake related destabilization of already unstable slopes. Obviously, pipelines can be damaged where they are buried in slide areas.

Pipelines can be pulled apart at joints, fail in bending, shear, tension, or compression or can be physically moved, changing their grade and making them inoperable for gravity flow.

8.3.3 Ground Motion

The behavior of buried pipelines subject to the seismic shaking has been observed to be controlled by the ground strains, predominantly in the axial direction (Kubo 1979). Continuous pipelines are put into tension and compression. If their ductility cannot resist this strain they can fail in tension. For continuous steel pipelines, local buckling (wrinkling) in compression is the expected failure mode. Where pipelines are discontinuous at bends, tees, or structure interfaces, failure in shear or bending may also occur.

To aid the design of buried lifelines, simplified analyses for the response of straight pipelines, bends and junctions to seismic ground waves are discussed. The major assumptions and limitations are that the ground has not failed nor liquefied.

8.3.3.1 Straight Pipelines

Analytical methods for straight pipelines can be grouped into two categories, namely: continuous and segmented pipelines. Steel pipelines with welded joints are an example of continuous pipelines. Ductile iron or PVC bell and spigot pipelines are examples of segmented pipelines. Continuous pipelines are generally modeled as beams with both axial and bending stiffness, which are connected to an elastic or elasto-plastic medium (i.e. the soil). Their responses can be determined in closed form. On the other hand, numerical solutions are typically required for the analysis of segmented pipelines.

8.3.3.1.1 Continuous Pipelines

The simplest method to analyze continuous pipelines is based on the assumption that buried pipelines are perfectly bonded to the surrounding soil (Sakuri 1969, Newmark 1967). In other words, it assumes no relative motion between the pipe and the ground. Thus, as upper bounds, one can take the seismic ground strains as the pipe strains and the seismic ground curvature as the pipe curvatures. This is equivalent to assuming that the pipe has no stiffness and therefore, will follow the ground exactly.

For the analysis and design of continuous pipelines, the upper bound of the axial strain of the pipe, ϵ_p, will be the maximum ground strain, ϵ_g, due to the earthquake:

$$\epsilon_p = \epsilon_g = \pm \frac{V_{max}}{C_p} \quad \ldots\ldots\ldots\ldots\ldots\ldots (1)$$

The upper bound for the maximum curvature of the pipeline X_p, will be the maximum ground curvature, X_g:

$$X_p = X_g = \frac{a_{max}}{C_s^2} \quad \ldots\ldots\ldots\ldots\ldots\ldots (2)$$

where V_{max} is the maximum ground velocity and a_{max} is the maximum ground acceleration during a seismic event at the site; C_p and C_s are the apparent longitudinal and transverse wave propagation velocities, respectively, of the controlling environments with respect to the pipeline.

If the site is subject primarily to body waves, C_p is the P-wave propagation velocity with respect to the ground surface and C_s is the S-wave propagation velocity with respect to the ground surface. If the site is subject to surface waves, C_p and C_s are the *Rayleigh* and *Love* wave propagation velocities.

If a pipeline with diameter D and the soil move together, the maximum pipeline bending strain ϵ_b is

$$\epsilon_b = \pm a_{max} \frac{D}{2C_s^2} \quad \ldots\ldots\ldots\ldots\ldots (3)$$

If a continuous piping system can meet the above sets of upper bound criteria (strain and curvature), the pipeline will be adequate against earthquakes that produce ground velocities and accelerations less than the V_{max} and a_{max} in an environment that has propagation velocities greater than the C_p and C_s used in the analysis.

For a soft soil site subject to surface waves, the ground strain, ϵ_g, given by equation (1) can be quite large and result in frictional slip at the soil pipe interface. In this case the pipe strain ϵ_p is a function of the frictional force at the soil pipe interface and the wave length of the surface wave. Analysis procedures by O'Rourke and Elhmadi (1990) are applicable in these cases.

For important projects such as nuclear power plants, the apparent wave propagation velocities must be investigated carefully. However, for preliminary design, the wave propagation velocities, C_s and C_p, for a site subject to body waves only may be determined by:

$$C_s = \sqrt{\frac{G}{\rho_s}} \text{ and } C_p = \sqrt{3}C_s \quad \ldots\ldots\ldots\ldots (4)$$

where G is the soil shear modulus and ρ_S the soil mass density.

For sites subject to body waves only, an apparent propagation velocity between 2000 and 3000 ft/sec (610 to 915 m/sec) can be used to yield conservative estimates of pipeline strain (ASCE 1983). Measurements of apparent propagation velocities of body waves by Japanese researchers (Harada 1983) indicate that the recommended values are indeed conservative and that a velocity of 3000 ft/sec (915 m/sec) would be appropriate as a lower bound estimate for these body wave situations.

If a substantial portion of the ground motion is due to surface (Rayleigh) wave propagation, the apparent propagation velocity could be as low as the shear wave velocity of the upper soil layer. In point of fact, some current Japanese procedures (Kubota Ltd. 1981) use that value for seismic design of buried pipe. Somewhat less conservative values for the apparent propagation velocity for surface wave situations are available in O'Rourke and Elhmadi (1988).

8.3.3.1.2 <u>Segmented Pipelines</u>

For segmented pipelines, the maximum relative joint displacements and maximum joint rotations become important design parameters in addition to the pipe strains and curvatures.

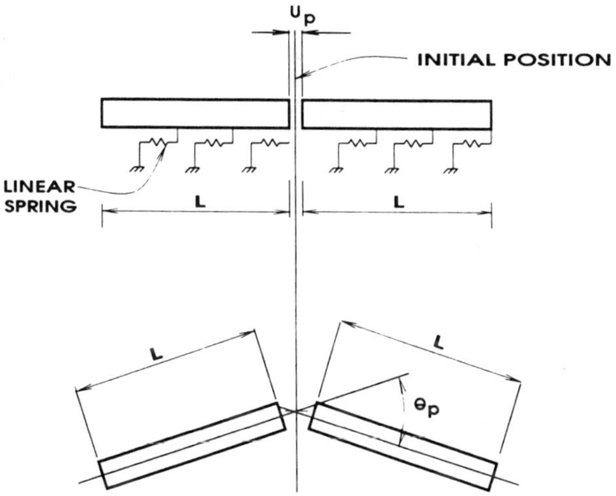

Figure 28
Static Joint Rotation

The simplest method for the analysis of segmented pipelines is the static solution (Wang 1979). It provides average values of the relative displacement and rotation in joints. Figure 28 shows the model considered in this method. The pipe segments are assumed to be infinitely rigid. It is also assumed that joint forces are negligible and the length of seismic waves exceeds significantly the dimension of the pipe segments. According to these assumptions, the average relative joint displacement U_p, and rotation θ_p are:

$$U_p = \epsilon_g L = \frac{V_{max} L}{C_p} \quad \ldots\ldots\ldots\ldots\ldots (5)$$

$$\Theta_p = X_g L = \frac{a_{max} L}{C_s^2} \quad \ldots\ldots\ldots\ldots\ldots (6)$$

where L is the length of the pipe segment.

Equations (5) and (6) are appropriate for straight segmented systems with flexible joints, such as ductile iron pipe with rubber gasketed joints. For straight segmented systems with stiff joints, such as cast iron pipe with lead caulked joints, equations (5) and (6) give the mean values for relative joint displacement and rotation. However, they significantly underestimate the maximum relative joint deformations for such stiff systems with realistic variations of joint stiffness from pipe segment to pipe segment. Using realistic variations of joint stiffness, O'Rourke and Elhmadi found that a few of the relative joint displacement for stiff joint systems were three to four times larger than the mean values given by equations (5) and (6). This facet is considered particularly important for those systems where a high damage rate would be one joint failure for every five hundred joints. This also partially explains why systems with flexible joints experience less seismic shaking damage than stiff joint systems.

8.3.3.2 Bends and Junctions

The methods available for the seismic analysis of bends and tees deal generally with continuous pipelines. In most cases, the seismic input is characterized by postulated ground displacement waves (Shah 1974, Goodling 1979). It is assumed that the pipelines are composed of elastic linear materials and that soils can be represented by rigid-plastic or elasto-plastic models.

8.3.3.2.1 Bends

Figure 29 shows the deformation of a bend when the direction of seismic wave propagation coincides with the pipeline referred to as element 1.

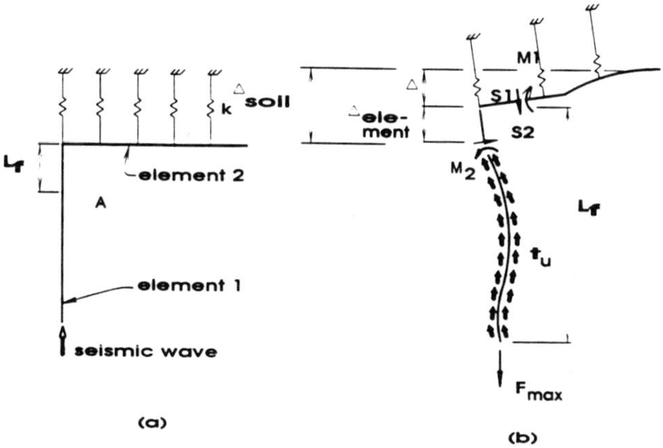

Figure 29
Force-Displacement at a Pipeline Bend L

At the bend as shown in Figure 29, the shears and moments (Shah 1974, Goodling 1979) are given as:

$$S_1 = \epsilon_g EA - t_u L_f \quad (7)$$
$$S_2 = S_1/3 \quad (8)$$
$$M = S_1/3\,\bar{\beta} \quad (9)$$

in which $\epsilon_g EA$ is the axial force induced in element 1 when there is no slippage and the soil strain is assumed to be ϵ_g, where ϵ_g can be determined from Equation (1) and EA is the axial stiffness of the pipe, $t_u L_f$ is the axial force induced in the same element over L_f when slippage occurs.

The maximum frictional force per unit pipe length t_u can be obtained from

$$t_u = \pi D\,(1 + K_o)\,\Upsilon\,H \tan \delta/2 \quad \ldots\ldots\ldots (10)$$

in which γ is the unit weight of soil, H is the depth of the pipe, K_o is the coefficient of earth pressure at rest, and δ is the interface friction angle between the soil and pipe.

$$L_f = \frac{4EA\bar{\beta}}{3K}\left(\sqrt{1+\frac{3\epsilon_g k}{2t_u\bar{\beta}}}-1\right) \quad \ldots\ldots\ldots\ldots (11)$$

in which

$$\bar{\beta} = \sqrt[4]{\frac{k}{4EI}} \quad \ldots\ldots\ldots\ldots\ldots\ldots (12)$$

I is the moment of inertia of the pipe, k = Dk_s, and k_s is the modulus of subgrade reaction.

8.3.3.2.2 Tees

Continuous tees can be analyzed by similar approaches. Figure 30 shows the deformation of a tee when the seismic wave travels along the element 1 without further explanation. The S_1 force and moment are given below:

$$S_1 = (\epsilon_g EA - t_u L_f)/2 \quad \ldots\ldots\ldots\ldots (13)$$

in which

and

$$L_f = \frac{EA\bar{\beta}}{2k}\left(\sqrt{1+\frac{4\epsilon_g k}{t_u\bar{\beta}}}-1\right) \quad \ldots\ldots\ldots\ldots (14)$$

$$M = S_1/2\bar{\beta} \quad \ldots\ldots\ldots\ldots\ldots\ldots (15)$$

The above mentioned static method of analysis for continuous pipe is simple but conservative. However, a quasi-static method (Wang 1979, 1983) is also available for a more refined analysis if needed.

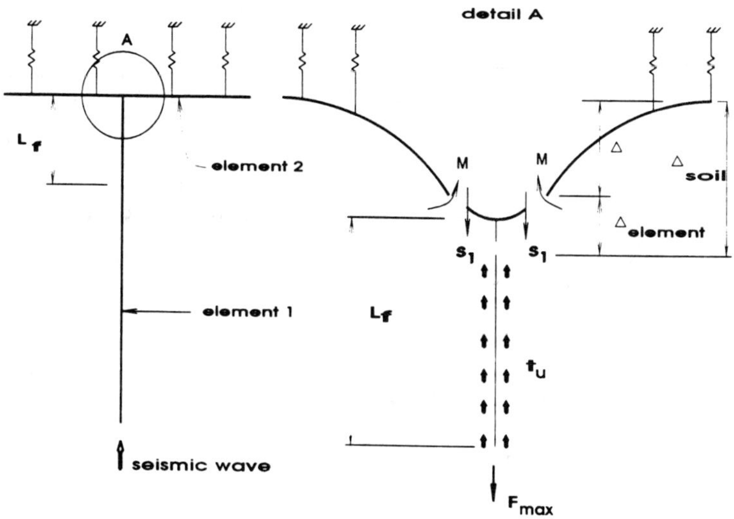

Figure 30
Force/Displacement at a Tee Junction

The seismic response of all types of segmented pipelines with bends and tees has not yet been determined. However, for 90 degree elbows in segmented systems with lead caulked joints, the total stress in the lead caulking of joints close to the elbow has been shown by O'Rourke and Bouabid (1989) to be significantly larger than the joints well removed from the elbow. In this regard, it should be noted that roughly two-thirds of the segmented pipeline breaks in the 1985 Mexico City earthquake occurred close to bends, tees, elbows, junction boxes, and other such "hard spots" (Ayala and O'Rourke, 1989).

8.3.3.3 Corrosion Related Failures

Failures resulting from pipeline corrosion can be accelerated by earthquake ground motion. The occurrence of small leaks has been known to dramatically increase for up to several years following an earthquake event.

8.3.4 Fault Crossings

Relative movement of a pipeline across a fault may be damaging. The analysis of pipeline response to ground movements requires input of soil displacement patterns and soil-pipeline interaction characteristics. The pipeline model may be based on continuous beam, jointed beam, shell, or combined beam and shell representations. Because soil and ductile pipe properties are governed by nonlinear stress-strain relationships, the refined analysis may require complex numerical procedures and are referenced herein.

Newmark and Hall (1975) were the first to propose a design method for a pipeline to resist large fault movement with several simple assumptions. Their method was later extended by Kennedy et al (1977) and Wang and Yeh (1985). Wang and Yeh's method (1985) is the most rigorous and most conservative, but requires complicated computer solution. Ford (1983) proposed the simplest method which could be unconservative because the friction forces between the pipe and the soil were neglected.

8.3.5 Exposed Structures - Above-ground Segments

Where it becomes necessary for a pipeline to be installed above-ground the supporting structure should be analyzed for seismic strength and stability. Site specific studies may be necessary depending on the nature and importance of the facility. The choice of material, for bridge or trestle structures could be influenced by the seismic characteristics of the materials and configuration.

8.3.6 Wall Penetrations

A critical point in pipeline continuity is at wall penetrations. Wall penetrations are found at vaults, pump stations, and storage structures. Pipes are typically restrained at wall penetrations and therefore require special designs. Possible designs include oversized holes with compressible seals to allow relative pipe/wall movement or articulated or expansion joints to allow pipe joint movement at the wall.

8.4 Seismic Resistant Design

8.4.1 Seismic Hazard and Geotechnical Assessment

Design earthquakes should be identified and the associated ground motion developed for each geologic area along the pipeline alignment. A geotechnical evaluation should be performed to locate areas where permanent ground deformation can be expected, such as, fault crossings and areas with high liquefaction susceptibility and landslide potential. In addition, identification of soil interfaces should be considered. The expected propagation velocity of seismic waves should also be determined.

8.4.2 System Redundancy

Redundancy should be built into water transmission and distribution systems. Multiple smaller pipes should be used in lieu of a single larger pipe to minimize reduction in operation due to breakage of pipes. Redundancy may also be applied to sewer force mains but is typically not applicable to tree structure sewer gravity collection systems except where pump stations and force mains are employed.

Isolation valves are required in the system to allow shut down of damaged pipeline sections in a redundant system. (See seismic or excess flow activated valves)

8.4.3 Pipeline Alignment

Liquefaction susceptibility, landslide potential, fault crossings, and levels of expected ground motion should be considered in pipeline alignment decisions.

8.4.4 Soil Stabilization

Soil stabilization should be considered as an alternative when considering pipeline alignments through soils with a high liquefaction susceptibility. Vibroflotation, drainage wells, and pressure grouting are alternative methods.

8.4.5 Pipe Material

Ductile pipe materials such as steel or ductile iron should be used to accomodate permanent pipeline deformations using continuous or restrained joint pipe. In general, pipes constructed with ductile material such as steel and ductile iron with restrained joints perform better than non-ductile material where permanent ground deformation is expected.

Segmented pipe with resilient joints can provide the ductility required to resist pipe damage where no permanent ground deformation is expected. Segmented pipe including ductile iron, plastics, steel and reinforced concrete with resilient gasket material can be used. Unreinforced concrete and vitrified clay with resilient gasket material is also acceptable.

8.4.6 Pipe Structural Design

Pipe tension, compression, bending, and shear loading must be taken into account in earthquake design of continuous pipelines in addition to non-earthquake loadings. As an alternative, stress can be relieved by designing flexibility into the system using flexible couplings, bellows couplings, or expansion loops.

8.4.7 Pipe Joint Design

For segmented pipelines, resilient gasketed joints should be used to provide for relative joint movements. For anticipated large ground movement in tension or compression, extra long restraining sleeves, multiple bellows couplings, or expansion loops should be used. To accommodate lateral offsets, flexible joints can be installed in series, or special combination ball joint/expansion joint fittings can be installed.

When feasible, shorter segments which will experience less strain imposed by the ground motion, should be used. Also, relative joint displacements are less for shorter segments.

Restrained joints should be used to keep the pipe spigot from pulling out of the bell where permanent ground deformation from liquifaction or landslide is expected.

8.4.8 Pipe Depth and Trench Backfill

Weight of backfill governed by pipe depth and backfill material weight determine resistance to pipe loading when subjected to differential ground movement. At fault crossings, shallow burial or above ground installations should be considered. Light weight backfill material should be taken into account.

8.4.9 Pipe Casings

Installation of pipelines inside a pipe casing at fault crossings may allow some fault offset with minimal pipe loading.

8.4.10 Appurtenances

8.4.10.1 Emergency Connections and Overflows

The system should be designed with the assumption that some earthquake damage will occur. Emergency connections with adjoining water purveyors should be considered.

Sewers should be designed to allow sewage overflows into receiving waters rather than allow the likelihood of backing up into basements or overflowing into streets.

8.4.10.2 Isolation Valves

Water system isolation valves should be installed to segregate pipelines with a high vulnerability from those with a lower vulnerability to earthquake damage. In the event of an earthquake, this will allow operators to close valves, segregating damaged portions of the system and more quickly restoring operation of the undamaged system. Valves should be periodically exercised and tested.

On large pipelines, careful consideration should be given to installing isolation valves that can be quickly closed (but not so quickly as to cause significant water hammer). One could then isolate pipe breaks that could themselves cause further damage from undermining or flooding.

8.4.10.3 Surge and Air Release/Vacuum (Air Inlet) Valves

Surge and/or air release valves should be considered to accommodate flows resulting from large line breaks that could damage the system. An example would be a large downstream break that could result in negative pressures upstream imploding the pipe or planned draining of the pipeline for maintenance.

8.4.10.4 Seismic or Excess Flow Activated Actuators

Seismic or excess flow activated valves should be considered which shut down systems that could be further damaged from earthquake induced pipeline leakage. Such a pipeline would then be inspected before putting it back on-line. Another application of this device is to minimize reservoir drainage through pipeline breaks.

The disadvantage of using these valves is that they can result in shutting off fire-flow which may be more important than leaking water. Their use must be weighed on a case by case basis.

8.5 Vulnerability Assessments and System Modeling

8.5.1 Vulnerability Assessments

A pipeline design should include considerations such as the nature and extent of vulnerability, the importance of the structure, and a discussion of functional outage on the entire system.

In many cases an all hazard type of review is appropriate so that a full range of hazards and risks can be discussed. In addition to the natural hazards of flood, landslide, and earthquake; factors such as sabotage or other induced situations should be considered.

8.5.2 System Modeling

With the widespread use of personal computers and available software, it may be appropriate to develop a system or network model for the pipeline segment being designed. The interrelationship of the segment being designed to the entire system can be modeled and sized more effectively once flow and operating perimeters are determined.

Seismic risk studies for lifeline systems were non-existent before the 1971 San Fernando earthquake. Since that event, much research has been conducted to develop the required databases and models for vulnerability studies. In general, the evolution of this type of modeling has proceeded from simple connectivity models to more complex connectivity/hydraulic modeling. Early studies demonstrated the usefulness of operation research type models for assessing connectivity of a water system after an earthquake.

More recent studies (Shinouzouka 1981,; Eguchi, 1981, 1983; and Taylor, 1986) have extended earlier models to include a subsequent model element that allows an analysis of water pressure and flow performance. This additional element is necessary in order to demonstrate the operation capability of the water system. Outage areas can occur where connectivity in the system has not been affected. In order to model the performance of a water system after an earthquake, a hydraulic analysis must be performed to identify low pressure and flow rate zones. This type of analysis has been done on systems owned by East Bay Municipal Utility District (Eguchi, 1983; Moghtaderizadeh, 1982), San Francisco (Auxiliary Water Supply) (Trautman, 1986), and Seattle (Ballantyne 1990).

Some of the benefits of performing a connectivity/hydraulic analysis are: 1) identification of seismically-vulnerable pipelines, 2) identification of potential water outage areas, 3) identification of possible repair levels after an earthquake, 4) development of emergency response plans, and 5) development of mitigation efforts.

In order to perform such an analysis, a particular water utility will require the following information:

1. Seismic hazard mapping including information on fault rupture, liquefaction, landslide, and ground motion. For some high risk areas, this information is available through state geological agencies or the U.S. Geological Survey.

2. System hydraulic network distribution models must be developed using programs such as KYPIPE (Woods 1981).

3. Seismic vulnerability models for individual water system components must then be selected

(Eguchi 1983, ATC-13 1985, ASCE-TCLEE 1990, Ballantyne 1990) or developed.

4. A computer program to estimate pipe failures and/or connectivity is required. This program will estimate failure probabilities for each pipeline link and each single site facility such as pump stations or storages given a particular earthquake scenario. Resulting flow rates and pressures are then calculated.

ASCE TCLEE (1991) has published a demonstration project that is designed to illustrate how a water utility can use available models and tools to estimate potential earthquake losses to its system. The emphasis is on economic loss and not system operation.

8.6 Emergency Response Planning

An emergency response plan, in-place before the earthquake, is a cost-effective mechanism for mitigating earthquake effects.

8.6.1 Earthquake Scenario

An earthquake scenario should be developed to be used as a planning tool for emergency response planning. The scenario should include a vulnerability assessment to estimate system losses.

8.6.2 Emergency Response Plan Development

Once system losses are estimated, procedures to respond to those losses should be developed. The plan should include procedures for secondary damage minimization, damage assessment, short term, and long term repair. For example, pipeline system plans should be widely distributed with additional copies kept in several locations not vulnerable to earthquake damage. Many of the pipeline system features described above will reduce damage and hence enhance emergency response.

8.6.2.1 Damage Assessment

Following a system interruption it is vital that accurate damage assessment be conducted. Accurate and timely information as to the nature and type of physical break as well as damage caused and immediate actions to be taken are all important first steps.

8.6.2.2 Repair Priority

In the situation where multiple incidents have occurred or widespread damage is being experienced, it is important to direct limited resources in the most affective way. System models as well as knowledgeable personnel need to have adequate direction to effectively restore a system. Normally, repair priority begins with the emergency backup facilities, then moves to the sources of supply and storage with distribution and individual customers following once supplies are restored. An important aspect of repair priority is knowing what parts of the system have been shut down (good documentation is necessary) so that the reverse process can occur effectively. Another use of the system models (Section 8.5.2) is a model that can be used in damage assessment and in repair prioritization.

8.6.2.3 Emergency Exercises

Emergency exercises are vital to the successful performance of personnel in a real emergency. Realistic emergency scenarios are needed for both tabletop and full scale exercises. Tabletop exercises are conducted with key, knowledgeable people that would actually be involved with an emergency working through a hypothetical problem and using mocked up communications links such as notes and room-to-room telephones.

Full-scale exercises are useful at least once a year involving the actual calling and responding of selected personnel and equipment in response to the emergency scenario.

8.6.3 Repair Equipment and Inventory

Equipment availability should be established before an earthquake event considering equipment available in-house as well as through mutual aid agreements and from local contractors.

Pipeline system operators should maintain an inventory of repair materials to be used in the immediate post earthquake period. For smaller pipelines, repair clamps and a nominal pipe inventory should be maintained. For larger diameter pipelines, pipe material that is available only through special order should be maintained near the pipeline alignment in an area not vulnerable to earthquake damage.

Trained personnel are required for system repairs. They may be available from within the organization or from an inventory of contractors, both local or outside the impacted area.

8.6.3.1 Personnel, Materials, and Equipment

The three elements of emergency response include personnel, materials, and equipment. The key ingredient in effective personnel use is communications. Multiple ways of communicating (telephone, radio, or other means), are needed. In addition, some agencies find that pocket size folded call-up cards are useful. "Call Up" call cards are useful in having the function, personnel name, radio, as well as office and home numbers. These cards need to be updated frequently so they are current.

Available materials include inventory on hand as well as materials available from local and regional suppliers. The ability to contact suppliers by means of secure phone lines, FAX, or other means should be discussed ahead of time and available to the agency.

Equipment during an emergency and qualified operators (personnel) can be a real problem. Since an emergency may tax the ability of the agency to respond, additional equipment and personnel will probably be needed. The contacts and phone numbers for local and regional equipment suppliers should be at hand. In addition, standard operating and stand-by rates should be agreed upon annually. Some agencies find simplified rental agreements helpful in this regard.

8.6.4 Replacement Costs

Replacement cost information is important in a situation where the reimbursement may be possible. In the situation where a federal or state emergency or disaster, may be declared, records will required of expenses and replacement cost information at the earliest time.

8.6.4.1 As-built Drawings/Construction Costs

The importance of as-built drawings to the repair and restoration process is recognized. Without knowing the type of construction and location in good detail it is difficult to affect repairs in a timely way. In addition, the month and year the project is built or constructed can be used with convenient indexes (such as the Engineering News Record or ENR Index) to update the replacement costs for a damaged facility. These original costs and replacement costs can be conveniently maintained once files are created for important, key facilities.

8.6.5 Relationships to Other Utilities

Most lifeline facilities are similar in their methods of operation. A common thread runs from one utility to another. Therefore an agency should consider maintaining contacts with other utilities on a regular basis. Therefore, should an emergency strike, communications between utilities can be facilitated. For example, should power be lost following an emergency, the restoration priority as to which area or pump station is most critical to the public need can be readily transmitted.

REFERENCES

1. Applied Technology Council, 1985, Earthquake Damage Evaluation Data for California, ATC-13, Redwood City, CA 94065.

2. ASCE Technical Council on Lifeline Earthquake Engineering, Advisory Notes on Lifeline Earthquake Engineering, 1983.

3. ASCE Technical Council on Lifeline Earthquake Engineering, Guidelines for Seismic Design of Oil and Gas Pipeline Systems, 1984.

4. ASCE Technical Council on Lifeline Earthquake Engineering, Seismic Loss Estimates for a Hypothetical Water System: A Demonstration Project, Craig E. Taylor, Editor, prepared by the Water and Sewage Committee and the Seismic Risk Committee of the American Society of Civil Engineers Technical Council on Lifeline Earthquake Engineering, ASCE, New York, 1990.

5. Ayala, G. and O'Rourke, M., "Effects of the 1985 Michoacan Earthquake on Water Systems and Other Buried Lifelines in Mexico", National Center for Earthquake Engineering Research, NCEER-89-0009, March 1989

6. Ballantyne, D.B., "Earthquake Damage Evaluation of Water and Sewage Lifeline Facilities, Loma Prieta Earthquake" NSF Grant Award BCS-9002459, Kennedy/Jenks/Chilton, Federal Way, Washington, December 1990

7. Ballantyne, D.B.; Taylor, C., "Earthquake Loss Estimation Modeling of the Seattle Water System", USGS Grant Award 14-08-0001-G1526, Kennedy/Jenks/Chilton, Federal Way, Washington, March 1990

8. Committee on Seismic Analysis, "Seismic Response of Buried Pipes and Structural Components," American Society of Civil Engineers, 1983.

9. Eguchi, R. T., Taylor, C. E., and Hasselman, T.K., 1983, "Earthquake Vulnerability Models for Water Supply Components," Redondo Beach, CA: J.H. Wiggins Company Technical Report No. 83-1396-2c, prepared for the National Science Foundation.

10. Eguchi, R.T.; Philipson, M.R.; Legg, J.H.; Wiggins, J.E.; and Slosson, J.E., "Earthquake Vulnerability of Water Supply Systems", Proc. Second Specialty Conference of Technical Council on Lifeline Earthquake Engineering, ASCE, Oakland, California, 1981

11. Elhamdi, K. and O'Rourke, M., "Seismic Wave Propagation Effects on Straight Jointed Buried Pipeline", National Center for Earthquake Engineering Research, technical Report NCEER-89-0022, State University of New York at Buffalo, August 1989

12. Ford, D.B., "Joint Design for Pipelines Subject to Large Ground Deformations," Earthquake Behavior and Safety of Oil and Gas Storage Facilities, Buried Pipelines and Equipment, PVP 77, ASME, June 1983, pp. 160-165.

13. Goodling, E.C. Jr., "Seismic Stresses in Buried Elbows," Presented at the April 2-6, 1979 American Society of Civil Engineers Convention and Exposition, held at Boston, Massachusetts (Preprint No. 3595).

14. Harada, T., Oda, T., and Shimada, T., "Determination of Apparent Wave Propagation Velocity from Damage Statistics for Buried Pipelines," Earthquake Behavior and Safety of Oil and Gas Storage Facilities, Buried Pipelines and Equipment, American Society of Mechanical Engineers, 1983, pp. 209-216.

15. Kennedy, R.P., Chow, A.W., and Williamson, R.A., "Fault Movement Effects on Buried Oil Pipeline," Transportation Engineering Journal of ASCE, Vol. 103, No. TE5, Sept. 1977, pp. 617-633

16. Kubo, K., Katayama T., and Ohashi, M., "Lifeline Earthquake Engineering in Japan," Journal of Technical Council, ASCE, April 1979, pp. 221-238.

REFERENCES - (Continued)

17. Kubota Ltd. "Earthquake Proof Design of Buried Pipeline", Kubota Ltd. Publication, Chuo-ku, Tokyo, Japan, October 1981

18. Moghtaderizadeh, M.; Wood, R.K.; Der Kiureghian, A.; and Barlow, R.E.; "Seismic Reliability of lifeline Networks", University of California, Berkeley, Journal of the Technical Councils of ASCE, Proceedings of the American Society of Civil Engineers, Vol. 108, No. TC!, May 1982.

19. Newmark, N.M., "Problems in Wave Propagation in Soil and Rock," Proceedings of International Symposium on Wave Propagation and Dynamic Properties of Earth Materials, Albuquerque, New Mexico, 1967, pp. 7-26.

20. Newmark, N.M., and Hall, W.J., "Pipeline Design to Resist Large Fault Displacement," Proc. U.S. Nat'l Conf. On Earthquake Engineering, Ann Arbor, MI, 1975, pp. 416-425.

21. O'Rourke, M. and Elhamdi, K., "Analysis of Continuous Buried Pipelines for Seismic Wave Effects", Journal Earthquake Engineering and Structural Dynamics, Vol. 16, 1988, pp 917-929

22. O'Rourke, M. and Ayala, G., "Seismic Damage to a Pipeline: A Case Study", Journal Transportation Engineering, ASCE, March, 1990

23. O'Rourke, M. and Bouabid, J., "Seismic Wave Propagation of Elbows in Buried Segmented Pipelines" Proc. Symposium Recent Developments in Lifeline Earthquake Engineering, ASME/JSME, Hawaii, July 1989

24. Sakurai, A. and Takahashi, T., "Dynamic Stresses of Underground Pipelines During Earthquakes," Proceedings of 4th World Conference on Earthquake Engineering, Chile, 1969, pp. 811-895.

25. Shah, H.H., and Chu, S.L. "Seismic Analysis of Underground Structural Elements," Journal of the Power Division 100, No. PO 1, American Society of Civil Engineers, July 1974, pp. 53-62.

26. Shinozouka, M.; Tan, R.Y.; and Toike, T., "Seismic Risk", Proc. Second Specialty Conference of Technical Council on Lifeline Earthquake Engineering, ASCE, Oakland, California, 1981

27. Taylor, C., Mc Donough, P.W., "Earthquake Risks to Utah Water and Gas Systems", Lifeline Seismic Risk Analysis - Case Studies, American Society of Civil Engineers, New York City, New York 10017-2398, 1986.

28. Trautmann, C.H., O'Rourke, T.D., Grigoriu, M., Khater, M.M., "Systems Model for Water Supply Following Earthquakes," Lifeline Seismic Risk Analysis - Case Studies, American Society of Civil Engineers, New York City, New York 10017-2398, 1986.

29. Wang, L.R.L., "A New Look Into the Performance of Water Pipeline Systems from the 1987 Whittier Narrows, California Earthquake", Old Dominion University Technical Report No. ODO-LEE-05, January 1990

30. Wang, L.R.L., and Cheng, K.M., "Seismic Response Behavior of Buried Pipelines," Journal of Pressure Vessel Technology, ASME, Vol. 101, No. 1, Feb. 1979, pp. 21-30.

31. Wang, L.R.L. and Olabimtan, A, "General Quasi-static Seismic Analysis of Buried Straight Piping Systems," Tech. Rept. LEE-006, School of Civil Eng. & Environmental Science, University of Oklahoma, Sept. 1983.

32. Wang, L.R.L., "Some Aspects of Seismic Resistant Design of Buried Pipelines", Proc. of ASME 3rd U.S. National Congress of Pressure Vessels and Piping, San Francisco, CA. June 25-29, 1979, PVP-34, pp. 117-131.

33. Wang, L.R.L., and Yeh, Y.H., "A Refined Seismic Analysis and Design of Buried Pipeline for Fault Movement," International Journal of Earthquake Engineering and Structural Dynamics, Vol. 13., No. 1., Jan-Feb. 1985, pp. 75-96.

34. Wood, Don J., 1980, Computer Analysis of Flow in Pipe Networks Including Extended Period Simulations, Program and User's Manual Department of Civil Engineering, College of Engineering of University of Kentucky, Lexington, Kentucky 40506.

APPENDIX I - NOTATION

The following symbols are used in this manual:

		Dimensions	
		English	Metric
A	area	sq ft	m^2
a	wave speed	fps	m/sec
c'	Coefficient depending on pipe restraint		
C_h	Hazen-Williams friction coefficient		
C_s	Scobey friction coefficient		
d (or D)	internal diameter of pipe	in. (or ft.)	cm (or m)
D/e	ratio of pipe diameter to wall thickness	Dimensionless	
DR	Dimension Ratio		
e	pipe wall thickness	in	cm
e	flow efficiency, actual Q/calculated Q x 100	%	%
E	Young's modulus of elasticity for pipe material	psi	kg/m^2
f	Darcy-Weisbach friction coefficient	Dimensionless	
g	gravitational acceleration constant	ft/sec^2	m/sec^2
h_f	energy loss due to friction in a length of pipe	ft	m
h_m	energy loss due to change in pipe alignment or appurtenance	ft	m
h_v	velocity head	ft	m
$H_{a,\ b...x}$	head (energy) with identifying subscripts	ft	m
K, K_L	minor loss coefficient	Dimensionless	
K	Bulk modulus of Elasticity	psi	kg/m^2

APPENDIX I - NOTATION (Continued)

l or L	length	ft	m
$N_{a,b...x}$	pump rotational speed	rpm	rpm
N_s	pump specific speed	rpm	rpm
n	Manning's or Kutters roughness factor		
n	speed ratio	Dimensionless	
$p_{1,2...n}$	pressure with identifying subscripts	psf or psi	kg/m^2 or kg/cm^2
Q	volume rate of flow	cfs	m^3/sec
R	hydraulic radius; equals d/r for pipes running full	ft	m
R	Reynolds number	Dimensionless	
S	energy loss per unit length of pipe	Dimensionless	
t_c	Critical time of valve closure	sec	sec
V, Vn etc.	mean velocity (Q/A) with or without identifying subscripts	fps	m/sec
$z_{1,2...n}$	elevation head with identifyingn subscripts	ft	m
γ	specific (unit) weight of fluid	lb/ft^3	kg/m^3
Δ (Delta)	difference, as: $\Delta h = h_2 - h_1$	ft	m
ϵ (epsilon)	equivalent roughness of pipe wall	ft	m
ϵ/d	relative roughness of pipe wall	Dimensionless	
μ (mu)	absolute viscosity	$lb\text{-}sec/ft^2$	$dyne\text{-}sec/cm^2$
μ	Poisson's ratio		
ν (nu)	kinematic viscosity	ft^2/sec	cm^2/sec
ρ (rho)	density (mass per unit volume)	$slug/ft^3$	gm/cm^3

APPENDIX I - NOTATION (Continued)

λ (lambda)	Poisson's ratio $\frac{\text{(lateral deformation)}}{\text{(longitudinal deformation)}}$	Dimensionless	
η	pump efficiency	%	%

APPENDIX II - ABBREVIATIONS

ANSI	American National Standards Institute, Inc. (formerly United States of America Standards Institute (USASI) and American Standards Association (ASA)
ASCE	American Society of Civil Engineers
ASTM	American Society for Testing and Materials
AWWA	American Water Works Association
b.e.p.	best efficiency point
BHP or bhp	brake horsepower
°C	degrees Celsius (formerly Centigrade)
cfs	cubic feet per second
cm	centimeter
cm^2	square centimeter
EL	Energy Line
°F	degrees Fahrenheit
FED	Federal Specification
ft	feet
ft^2	square feet
ft^3	cubic feet
fps	feet per second
gm	gram
gpm	gallons per minute
HGL	Hydraulic Grade Line
HP or hp	horsepower
hr	hour
ID	internal, or inside, diameter of pipe
in.	inch
kg	kilogram
km	kilometer
kw	kilowatt
KL	Loss Coefficient
lb	pound
lin ft	linear foot
m	meter
m^2	square meter
m^3	cubic meter

mg	milligram
mgd	million gallons per day
mg/l	milligrams per liter
min	minute of time
mm	millimeter
mps	meters per second
NPSH	net positive suction head
PRV	Pressure Reducing Valve
psf	pounds per square foot
psi	pounds per square inch
Ref.	Reference in Appendix 1
rev	revolutions
rpm	revolutions per minute
sec	second of time
sec^2	second square
sq ft	square foot
WHP	water horsepower
yr(s)	year(s)

INDEX